PLOTTING FROM ABOVE

ENHANCING AGRICULTURAL MAPPING IN ASIA AND THE PACIFIC

Anthony Burgard; Anna Christine Durante; Pamela Lapitan;
Mahinthan Joseph Mariasingham; Arturo Y. Pacificador, Jr.; and Mashal Riaz

JUNE 2024

ASIAN DEVELOPMENT BANK

ADB

Notes:
In this publication, "$" refers to United States dollars.
ADB recognizes "Republic of Armenia" as Armenia.
Photos: All photos by the Asian Development Bank unless otherwise indicated.

Featured on the cover, starting from the top left and moving clockwise, are images capturing fieldwork conducted in Armenia, the Cook Islands (including the third photo), and the Lao People's Democratic Republic. The fifth photo is from a remote sensing training session in Viet Nam.

CONTENTS

TABLES AND FIGURES

FOREWORD

This report *Plotting from Above: Enhancing Agricultural Mapping in Asia and the Pacific* provides a comprehensive overview of the application of a methodology for agricultural area measurement and insights gained from its implementation in these three countries.

The Asian Development Bank (ADB) has launched a technical assistance project to strengthen the capabilities of national statistics offices and other ministries, equipping them with the necessary skills to meet the Sustainable Development Goals' increasing data demands. A pioneering aspect of this project is the use of geospatial technologies, which have been employed to create methodological tools for agricultural area measurement. These tools are designed to evaluate the discrepancies in agricultural area estimates between traditional farmer-reported methods and more modern approaches using Global Positioning System (GPS) devices for objective measurements.

Agricultural land is a crucial asset for farmers, serving as the foundation of their economic livelihood. It facilitates various activities such as crop cultivation, animal husbandry, fisheries, and forestry. Historically, obtaining accurate and unbiased measurements of agricultural lands has been a challenging aspect of agricultural statistics. However, by identifying and addressing biases in these measurements, policymakers can gain a more precise understanding of agricultural productivity. The advent of geospatial technologies has made this task more accessible and economical, revolutionizing the way agricultural areas are measured. This technological shift has made objective area measurement not only more feasible but also more cost-effective.

This report features an in-depth analysis of an area frame approach implemented in the Cook Islands. The approach uses non-overlapping and relatively fixed geographical units from which a sample may be drawn, instead of the more time-consuming traditional list frame comprising agricultural holdings compiled during an agricultural census. Geospatial data can be integrated with a sample of polygons drawn from an area frame, which can be further stratified by various characteristics such as topography.

Additionally, the report delves into the use of geospatial technologies to assess biases in agricultural land reporting in Armenia, the Cook Islands, and the Lao People's Democratic Republic (PDR). The study employed a range of geospatial techniques for an unbiased measurement of agricultural land. One of the key strengths of this study is its exploration of the feasibility and applicability of the approach across three countries in different regions with distinct characteristics, agroclimatic conditions, and socio-political contexts.

The ADB project team would like to take this opportunity to thank the implementing agencies from the Statistical Committee of the Republic of Armenia, the Cook Islands Ministry of Agriculture, and the Lao PDR Ministry of Agriculture and Forestry for their invaluable contributions and offering critical insights into the diverse agricultural practices across Asia and the Pacific. The team is grateful to Mr. Gagik Ananyan, Deputy of the President of the Statistical Committee of the Republic of Armenia; Mrs. Temarama Anguna-Kamana, Secretary of Agriculture, Cook Islands Ministry of Agriculture; and Ms. Khamvay Nanthavong, Director of the Center for Agricultural Statistics of the Lao PDR Ministry of Agriculture and Forestry for spearheading the implementation of the project in the pilot areas. Sincere gratitude is offered to Arsen Avagyan, William Wigmore, Tearoa Iorangi, Puna Kamoe, Angeylie Ngaoire, and Sengphachan Khounthikoummane, for providing technical and logistical support in the conduct of all project activities. Further, the team would like to express appreciation to field and technical staff for their dedication to collecting, processing, and analyzing the data from which this study would not be possible.

The report has been produced by the Statistics and Data Innovation Unit within the Economic Research and Development Impact Department at ADB, under the overall direction of Elaine S. Tan. The project and report teams were led by Mahinthan Joseph Mariasingham, with valuable research and technical support from Anthony Burgard, Anna Christine Durante, Pamela Lapitan, Arturo Y. Pacificador Jr., and Mashal Riaz. Melanie Kelleher copyedited the final manuscript, Edith Creus typeset the report, and Claudette Rodrigo prepared the cover design.

It is hoped that this study aids in the evolution of methodologies for measuring agricultural areas. This methodological advancement holds the potential to enhance accessibility and accuracy in agricultural data collection. It is anticipated that this report will have a positive influence on the future of agricultural statistics.

Albert Francis Park
Chief Economist and Director General
Economic Research and Development Impact Department
Asian Development Bank

INTRODUCTION

A Technological Shift in Agricultural Statistics

The growing accessibility of geospatial technologies is reshaping how agricultural statistics are gathered, processed, and disseminated. Advanced technologies like remote sensing using satellite imagery, Global Positioning System (GPS), and unmanned aerial vehicles (UAVs) offer the potential for more efficient methods to monitor changes in agriculture with greater precision and frequency.

Once cost-prohibitive for large-scale statistical data collection, geospatial technologies are becoming increasingly commonplace, efficient, and accessible in official statistics. Consumer-grade equipment is now more capable and less expensive. As computer-assisted interviewing data collection methods once ushered in the digitization of statistical data collections, the same modern technologies in tablet computers provide a means to simplify geospatial data collection. It is now more commonplace for agricultural data producers to regularly collect, for example, the point location of agricultural households and boundary areas of agriculture and croplands.

Similarly, the cost of remote sensing satellite imagery has decreased, largely due to initiatives by organizations like the European Space Agency (ESA), the Japan Aerospace Exploration Agency, and the National Aeronautics and Space Administration, which have made high-resolution satellite imagery more readily available as a free global public good.

Greater accessibility of UAVs and ultra-high-resolution imagery have empowered, for example, the Pacific island countries—which have traditionally been susceptible to climate change—to regularly map and monitor land changes with greater detail and timeliness. These advancements pave the way for the broader adoption and utilization of geospatial technologies, significantly enhancing agricultural statistics.

One of the objectives of the Asian Development Bank (ADB) technical assistance project is to integrate geospatial technologies into traditional survey design and build the capacity of data producers in the region on its potential use case applications. Despite the increasing relevance of such technology, national statistical offices and official producers of agricultural statistics often face challenges in its utilization. This includes hiring staff skilled in working with these data types and upgrading information technology infrastructure to manage the larger datasets. While many organizations have established geographic information system (GIS) units within their offices, their application is primarily limited to creating and distributing maps. There has been limited progress by comparison in integrating geospatial techniques into survey design and data collection processes.

One innovation, however, uses geospatial techniques for survey design and improves upon the traditional sampling list frame. Construction of the traditional list frame—a comprehensive list of agricultural holdings in a country—is a costly undertaking often completed once every 5 to 10 years during agriculture censuses and—if well maintained—updated through a system of intercensal agricultural surveys. The list frame is a pain point for many official data producers as it is notoriously difficult and expensive to build, update, and maintain. While countries in the region have trended to pursue a farmer-registry-based approach to the construction of these frames, they are currently limited by low farmer volunteer rates and may lack key auxiliary data to serve as an effective sampling list frame. A complementary approach has been for official statistics producers to adopt a mixed-frame approach incorporating the use of an area frame. An area frame consists of non-overlapping and relatively fixed geographical units from which a sample may be taken.

A limitation of area frames is that they are not always optimized for statistical purposes and are constructed with broader use cases, such as the demarcation of administrative boundaries. This often results in insufficient auxiliary information to enhance survey sample efficiency. With the growing accessibility of geospatial data, opportunities exist to integrate diverse geospatial data sources, including land use or land cover maps, topographical maps, and the locations of natural and human-made features. Using location as a reference point, these data can be integrated to enhance sampling methods. This integration allows for more efficient or "smart" sampling approaches—such as stratifying by differing agricultural characteristics—thereby improving the accuracy and efficiency of data collection processes. This stratification could, for example, be based on the density of the estimated agricultural area or distinct agroecological zones, thus improving the precision of the survey sample.

In several Asian countries, national statistical offices and agricultural line ministries are transitioning to precision agriculture using digital records and geospatial information to map agricultural areas. This includes digitizing agricultural parcels to enhance production statistics estimation as seen in the last agriculture census of the People's Republic of China, the Smart Farm Initiative in the Republic of Korea, and the agricultural administrative record system in Sri Lanka. Additionally, these platforms offer the potential for a system of ground truth validation points required for other technical estimation such as remote sensing-based crop estimation.

This paper will explore a case study for implementing an area frame approach in the Cook Islands using a land cover map developed with the assistance of ESA as an auxiliary data source. The land cover map—created from high-resolution satellite imagery of Rarotonga Island in 2021—enabled the classification of areas with a high probability of agricultural production. This classification was instrumental in determining the sample allocation for a post enumeration survey conducted after the 2021 Agriculture Census.

The paper will further investigate using geospatial technologies to evaluate biases in reporting agricultural land in Armenia, the Cook Islands, and the Lao People's Democratic Republic (Lao PDR). The study applied diverse geospatial methods for an objective measurement of agricultural land, including specialized handheld GPS devices and tablet-based software for digitizing the boundaries of agricultural parcels on high-resolution satellite imagery.

Agricultural Land is a Key Factor in Economic Production

Agricultural land is a key productive asset for farmers, forming the base of their economic livelihood. It is a key factor of production, enabling the growing of crops, raising animals, fisheries, and forestry activities. The size of agricultural land is a critical statistic from a policy perspective as it helps policymakers better understand how farming is structured in a country. Accurate measurements of agricultural area are important for evaluating how productive farms are, planning for agricultural growth, and making effective agricultural policies.

From a macroeconomic perspective, agricultural land area feeds into the critical calculation of potential economic output. In many cases, crop production statistics are derived based on reported agricultural area multiplied by an average yield estimate for the locality. Any biases in land area estimates reported by the farmer will significantly compound the estimate for total agricultural output.

Biases in estimating agricultural areas may also negatively impact resource allocations to government support programs. Food security is a primary concern for many governments, especially those in climate-vulnerable areas such as the Pacific island countries. An overestimation or underestimation of agricultural land area—and thereby production—can affect the ability of a country to meet its caloric and nutrient needs sustainably. And where food security is a challenge, improved statistics on the agricultural area help governments improve plans to import key agricultural commodities to meet these requirements. These data may then be used to support crop and input subsidies, crop insurance schemes, and additional technical support through agricultural extension services.

Finally, agriculture significantly impacts the environment, contributing to greenhouse gas emissions and encroachment on natural forests, leading to biodiversity loss. Improved estimates of agricultural land use are essential for urban and rural land use planning and managing these impacts in the long term.

Recall Biases in Estimating Agricultural Land

The common method for estimating agricultural land in census and surveys involves subjective recall by farmers, asking them to report the extent of their operated land. This approach presents several challenges. It assumes farmers' understanding of the term operational agricultural land, a concept that can introduce non-sampling biases if unclear. These issues are compounded by differences in agricultural practices throughout the region, where in some countries, it may be common to share communal areas for agricultural production and live and work in areas where land tenure is not clearly defined.

This method assumes farmers have accurate knowledge of the size of their land. This knowledge is typically based on information from formal land titles or deeds, providing official documentation of precise measurements of the land. Without these documents, farmers' estimates may be largely speculative. In Armenia, for example, a strong land cadastral system exists in which land area is linked with property tax and government planning systems. In such cases, farmer knowledge of their area is based on how updated these systems are and the extent of public access to these documents.

There is also the issue of reported area measurement units. Familiarity with standard units of area measurement is not always common among farmers. For instance, in the Cook Islands, a farmer might describe their land in terms of a rugby field size or reference natural landmarks like a large tree. In the Lao PDR, local units such as "*lai*" and "*ngam*" are common, complicating accurate reporting in standardized units like acres or hectares. Farmers may use different area units for different land features in certain instances. For example, total area might be reported in acres, while individual plots are described in square meters (m^2). Differences in area units reported may burden the field staff when performing quality assurance checks.

The lack of familiarity with standardized units of measurement adds to the response burden for farmers, leading them to either inaccurately convert the units or give speculative answers. To address this, agricultural surveys increasingly permit respondents to specify the units used for each area-related question. However, inconsistencies in reported areas remain a common issue. These require careful validation and adjustment during the data processing to ensure accuracy and consistency.

Finally, the physical characteristics of the land can hinder accurate area estimation. In many Asian and Pacific countries—especially smallholder mixed cropping systems—agricultural parcels often have irregular shapes. They may be situated in steep mountainous terrain, affecting farmers' perception of their size. This complexity can lead to inaccuracies in reported land area, significantly impacting the quality of agricultural statistics.

Introducing Objective Measurements to Assess Agricultural Land

Acknowledging the subjective biases associated with farmer recall data, one approach to mitigate this is by introducing objective area measurement. Traditionally, objective area measurement involves land surveying to delineate agricultural parcels using the tape-and-compass approach (FAO, 1982). This method entails measuring each side of the parcel with a tape measure and using a compass to determine the angles between sides. However, this technique is time-consuming and resource-intensive. It also demands highly skilled workers to perform precise measurements and calculate the area using complex trigonometric functions. When done correctly, however, the tape and compass method is considered the "gold standard" for agricultural area measurement (Carletto et al., 2016).

In 2018, ADB conducted a pilot study to explore using GPS and satellite data as technological alternatives for objectively measuring agricultural parcels. The study findings indicated that these methods aligned well with the "gold standard" of measurement and, on average, were more cost-effective to implement (Dillon et al., 2018). The cost-effectiveness of GPS technology in land area measurement is primarily due to its streamlined implementation in the field. The method involves an enumerator physically (usually led by the farmer for guidance) walking the perimeter of a parcel while carrying a GPS device, which automatically tracks and calculates the area. This approach reduces the need for multiple pieces of surveying equipment and extensive training of field staff. Furthermore, the time required for measurement is limited to the duration of the perimeter of the parcel, significantly streamlining the overall process.

Another popularized area measurement technique uses satellite imagery to digitize or trace parcel boundaries on a tablet computer remotely. This technique enhances efficiency by physically eliminating the need to walk the perimeter of the parcel. However, this method faces challenges. First, it requires farmers to accurately identify their land from an aerial perspective and be able to recognize parcel boundaries using identifiable landmarks in the image. Additionally, agricultural areas are dynamic, often changing between seasons within the same year. Therefore, current and up-to-date imagery is crucial to avoid discrepancies from outdated images. With adequate planning and the growing availability of timely, high-resolution satellite imagery, this method has the potential to be effective. It is particularly viable in countries where parcels are physically large to efficiently traverse or challenging due to environmental hazards.

The efficacy of land measurement methods can significantly vary across diverse geographic areas, as demonstrated in this study that examines their implementation in three distinct countries: Armenia, the Cook Islands, and the Lao PDR. Armenia features large, structured, mono-cropped fields. In contrast, the Cook Islands—characterized by smallholder farming within an island context—has notably smaller parcel sizes. The Lao PDR presents a mix of large mono cropped areas and environmental challenges from its highly mountainous terrain and access difficulties.

METHODOLOGY

Study Area

The study was conducted in three study areas: Yerevan (Armenia), Rarotonga (Cook Islands), and Vientiane (Lao PDR). The selection of these countries aimed to enhance the capabilities of National Statistics Offices to meet the data demands of the Sustainable Development Goals (SDGs) indicator framework, more specifically by employing innovative technologies and methodologies for sampling frames.

In the Cook Islands, the area sampling frame and the area measurement methodology were employed in a representative survey for Rarotonga Island. Meanwhile, only the area measurement methodology was pilot-tested in Armenia and the Lao PDR at the village and district levels.

In Rarotonga, Cook Islands, the methodology employed by this study was used as part of the 2022 Post Enumeration Survey (PES) for the Agriculture Census of the same year. The PES aimed to (i) apply a dual-frame survey design using land cover maps developed for the reference year, (ii) evaluate coverage errors arising from varying definitions of an agricultural holding, and (iii) compare farmer-reported land areas with objective GPS measurements to identify biases.

The geographic scope of the study encompassed Rarotonga Island in the Cook Islands, with a reference period set for the 2021 calendar year. The scope of the study included 78 enumeration areas on Rarotonga Island, where 2,286 households were identified as agricultural based on the 2021 Census of Population and Dwellings (COPD). Interviews were conducted with all agricultural holdings within each selected enumeration area, targeting a sample size of 426.

Rarotonga—the largest island in the Cook Islands—covers 67 square kilometers (km²) and houses about 10,000 people. In 2011, it hosted 40% of the nation's agricultural households, mainly engaging in small-scale and subsistence farming of taro; other root crops; and fruit trees like pawpaw, banana, and citrus. Agriculture primarily occurs in the lowland peripheries of the island, with some rare plant species harvested from forested or wild areas inland. Agriculture by nature of the prominent crops tend to be harvested continuously throughout the year.

In Yerevan, Armenia, the methodology was implemented as a pilot in Berdik village (part of Artashat Town) to evaluate discrepancies between land area measurements reported by farmers and those obtained via GPS.

Berdik village—situated approximately 30 minutes from the capital—is an agricultural community with 180–200 households predominantly producing grapes and fruit orchards, including apricots, apples, cherries, and plums. Many households also process food products like jams and alcoholic liqueurs. The village has a well-delineated official land cadastre system that records the area measurements for each agricultural land block.

Farmers are also issued official land deeds with the dimensions of their lands. Despite the presence of this cadastral system, it was observed that the last official measurements were taken 15 years previously. Over time, some farmers reorganized or combined these land blocks, adjusting to changes in their production needs.

In Vientiane Province, Lao PDR, the methodology was tested in the Pak Pok village in Vang Vieng district to assess land area measurement biases between farmer reports and GPS data. Vang Vieng district—approximately a 3-hour drive from the capital—is renowned for its rice production, which predominantly occurs in the plains but is shaped by the surrounding mountainous valley. In Pak Pok village, the agricultural holding structure is unique as farmers often reside in separate dwellings from their agricultural land, commuting to their parcels for agricultural work. For this study, interview appointments were arranged with the local village head to meet farmers directly at their agricultural parcels.

Study Description

The following section outlines the methodologies applied in Armenia, the Cook Islands, and the Lao PDR for this study. These sections will elaborate on the specific methodologies and tools employed for data collection and assess the insights and practices gained from the study.

Cook Islands

Compilation of the area frame for the 2022 Post Enumeration Survey

The primary objective of the 2022 PES involved implementing an area frame approach in the survey design. To achieve this, ADB collaborated with ESA to estimate a land cover map for the islands of Rarotonga and Aitutaki. Figure 1 shows the land use and land cover map generated by ESA. This map was developed utilizing Worldview-2 satellite imagery (resolution 0.46m) and a classification system that aligns with the United Nations Land Classification System. It distinctly categorizes the land into 12 different classes. The land classification was estimated on satellite imagery exactly 1 year before the start of the PES data collection to ensure no variance due to the differences in seasons.

To enhance the survey design of the 2022 PES using the land cover map, the study considered consolidating the United Nations Land Classification System categories into two broad groups: agriculture and non-agriculture. In this process, agriculture was defined as a combination of classes A12–A73, and B11, as referred to in the legend of Figure 1.

After the aggregation of classes, the study calculated the agricultural intensity for each enumeration area. Agricultural intensity is the percentage of agricultural land relative to the total land area (Figure 2).

The sample design employed a one-stage cluster sampling methodology (Appendix 1: 2022 Cook Islands PES Survey Design). Enumeration areas were selected from three strata, each classified by the intensity of the agricultural area. This classification was based on land cover maps produced in collaboration with ESA. For the sample selection, 12 enumeration areas were chosen across the three strata, with an equal distribution of four from each stratum.

Figure 1: Land Use and Land Cover Classifications of Rarotonga

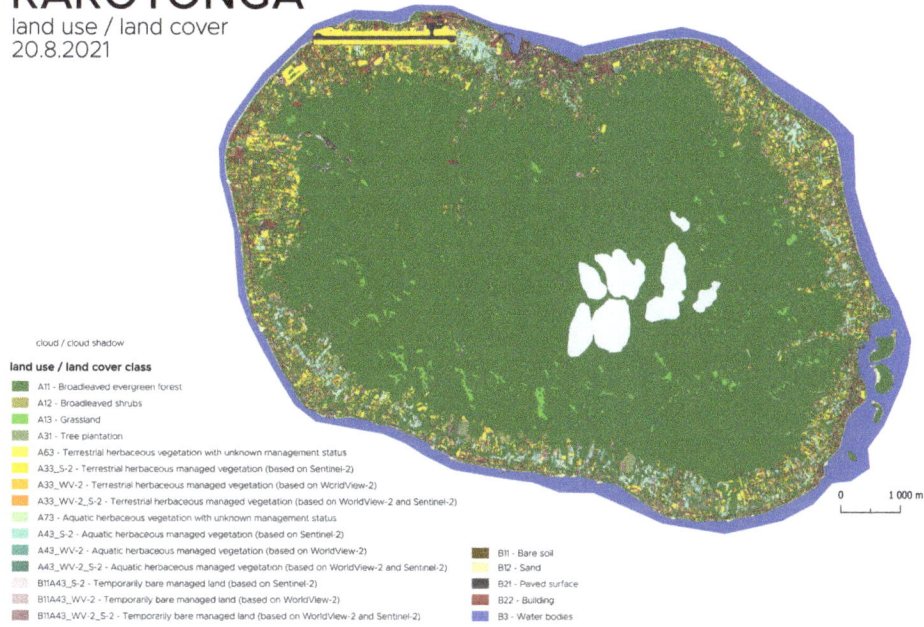

RAROTONGA
land use / land cover
20.8.2021

cloud / cloud shadow

land use / land cover class
- A11 - Broadleaved evergreen forest
- A12 - Broadleaved shrubs
- A13 - Grassland
- A31 - Tree plantation
- A63 - Terrestrial herbaceous vegetation with unknown management status
- A33_S-2 - Terrestrial herbaceous managed vegetation (based on Sentinel-2)
- A33_WV-2 - Terrestrial herbaceous managed vegetation (based on WorldView-2)
- A33_WV-2_S-2 - Terrestrial herbaceous managed vegetation (based on WorldView-2 and Sentinel-2)
- A73 - Aquatic herbaceous vegetation with unknown management status
- A43_S-2 - Aquatic herbaceous managed vegetation (based on Sentinel-2)
- A43_WV-2 - Aquatic herbaceous managed vegetation (based on WorldView-2)
- A43_WV-2_S-2 - Aquatic herbaceous managed vegetation (based on WorldView-2 and Sentinel-2)
- B11A43_S-2 - Temporarily bare managed land (based on Sentinel-2)
- B11A43_WV-2 - Temporarily bare managed land (based on WorldView-2)
- B11A43_WV-2_S-2 - Temporarily bare managed land (based on WorldView-2 and Sentinel-2)

- B11 - Bare soil
- B12 - Sand
- B21 - Paved surface
- B22 - Building
- B3 - Water bodies

0   1 000 m

Source: European Space Agency.

Figure 2: Agricultural Intensity on Rarotonga Island, Cook Islands

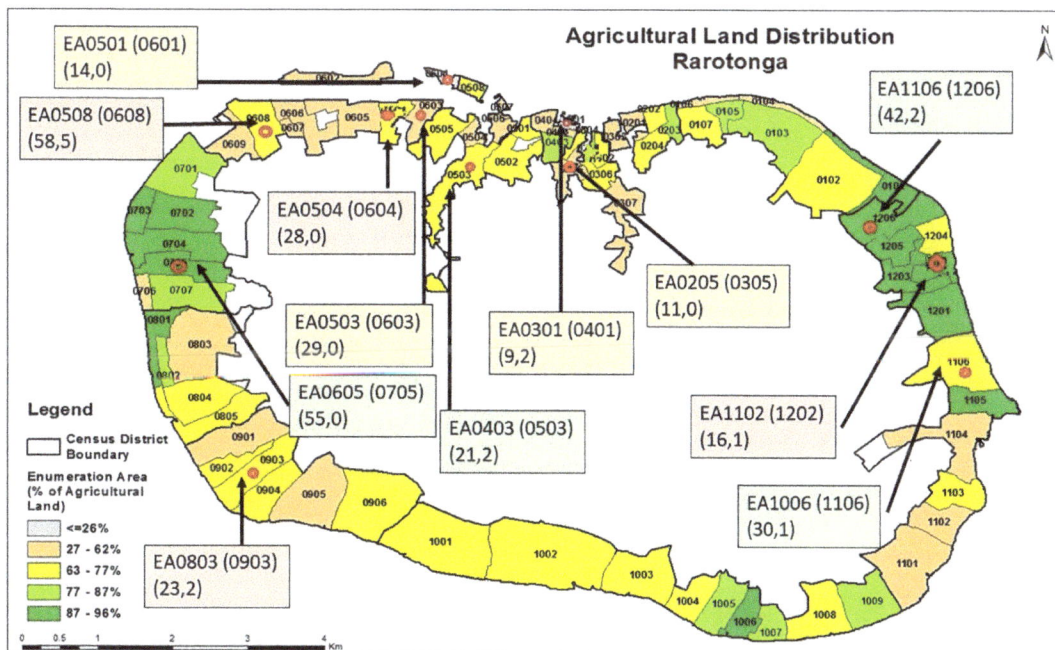

Agricultural Land Distribution Rarotonga

EA0501 (0601) (14,0)

EA0508 (0608) (58,5)

EA1106 (1206) (42,2)

EA0504 (0604) (28,0)

EA0205 (0305) (11,0)

EA0503 (0603) (29,0)

EA0301 (0401) (9,2)

EA0605 (0705) (55,0)

EA1102 (1202) (16,1)

EA0403 (0503) (21,2)

EA1006 (1106) (30,1)

EA0803 (0903) (23,2)

Legend

Census District Boundary

Enumeration Area (% of Agricultural Land)
- <=26%
- 27 - 62%
- 63 - 77%
- 77 - 87%
- 87 - 96%

0 0.5 1   2   3   4 Km

Source: Asian Development Bank estimates.

The stratification for the study was delineated according to the potential agricultural area: Stratum 1 comprised areas smaller than 10.93 hectares; Stratum 2 included areas between 10.93 and 31.18 hectares; and Stratum 3 consisted of areas larger than 31.18 hectares. These strata are designed to represent the "small," "medium," and "large" enumeration areas in terms of the potential area of agriculture as derived in Figure 2. A selection of three strata was chosen to be most optimal considering the total number of enumeration areas of 78. The stratum boundaries were derived using the cumulative square root of frequency (Cum √F), established by Dalenius and Hodges (1959), where the aim is to minimize variability within the strata and maximize variability between them.

The use of the land cover map and its classifications to improve the allocation of samples is an example of how geospatial information can be used to improve the efficiency of the sample by allocating samples toward areas where agricultural activity is likely to be. This strategy is particularly beneficial when the geographic spread of agricultural land is not well-documented, contributing to a more unbiased sample selection by ensuring coverage is informed by the land cover map.

Nonetheless, the approach has its limitations. There must be a reasonable degree of confidence in the accuracy of the land cover map, as it is a modeled estimate with inherent biases and assumptions. Land cover classification accuracy is gauged against ground truth data from current high-resolution satellite imagery, land cover maps in 2009, and the 2011 Agricultural Census. The overall classification accuracy is calculated as the proportion of correctly classified instances out of the total observations (n=114). Beyond the general accuracy, it is notable that 85% of the actual area of the vegetation class was correctly identified.

Understanding the accuracy and biases of a classification is particularly relevant given that the calculation of agricultural intensity relies on further class aggregates based on choices of the survey designer and government consultations. It is crucial to align the land cover map's time period with the data collection period. Since agricultural intensity fluctuates with seasons, discrepancies can result in inefficiency in sample allocation. For example, it may be common to have misclassifications where open sports fields or public parks can be misconstrued for agricultural areas.

Assessing Coverage Error through the 2022 Post Enumeration Survey Cook Islands

Following the selection of enumeration areas, the 2022 PES Cook Islands conducted interviews with all households listed in the 2021 COPD deemed engaged in agriculture. Considering the differences in agricultural definitions between the COPD and the Ministry of Agriculture (MOA), this approach enabled the survey to reapply the definition of the MOA to estimate the count of agricultural households and assess the extent of agricultural activities on Rarotonga Island.

As a baseline, data from the 2021 COPD for all the Cook Islands showed a decrease in agricultural households from 2,098 in 2011 to 1,851 in 2021, with the decline in subsistence-only households offset by increased commercial agricultural producers (Table 1).

Table 1: Number of Agricultural Households by Level of Agricultural Activity from 2011 and 2021 Censuses of Population and Dwelling Cook Islands

Level of Agricultural Activity	2021 COPD		2011 COPD	
	Frequency	%	Frequency	%
Subsistence Only	1,590	85.9	1,860	88.7
Subsistence and Commercial	225	12.2	215	10.2
Commercial Producers	36	1.9	23	1.1
Total Number of Agricultural HH	1,851		2,098	

COPD = Census of Population and Dwelling, HH = households.
Source: Asian Development Bank estimates based on the Cook Islands Census of Agriculture 2011.

In the 2021 COPD, an agricultural household is identified by engagement in agricultural activities, including livestock raising, crop and root crop cultivation, and vegetable and ornamental plant growing. Conversely, the 2011 and 2021 agricultural censuses apply specific minimum threshold criteria to define an agricultural household (Figure 3). Discrepancies in the number of agricultural households reported by the 2011 COPD and the 2011 Census of Agriculture (COA) can mainly be attributed to differing definitions of what constitutes an agricultural holding. Differences in scope may be partially mitigated since the 2011 Agriculture Census frame is derived from the 2011 COPD.

Acknowledging the differences in how an agricultural household is defined between the population and agriculture censuses, the 2022 PES estimates an increase in the number of agricultural households, compared to the figures from the 2011 COA, driven by the increase in minor agricultural households. In contrast, there was a significant reduction in the estimated numbers of both subsistence-only and commercial households in the 2022 PES (Table 2).

Moreover, employing the number of agricultural households from the 2021 COPD as an auxiliary variable in the ratio estimator resulted in lower coefficient of variation values compared to the expansion estimator. This demonstrates the potential of improving estimates by utilizing COPD data.

Figure 3: Definitions for Level of Agricultural Activity in 2011 Census of Agriculture and the 2022 Post Enumeration Survey, Cook Islands

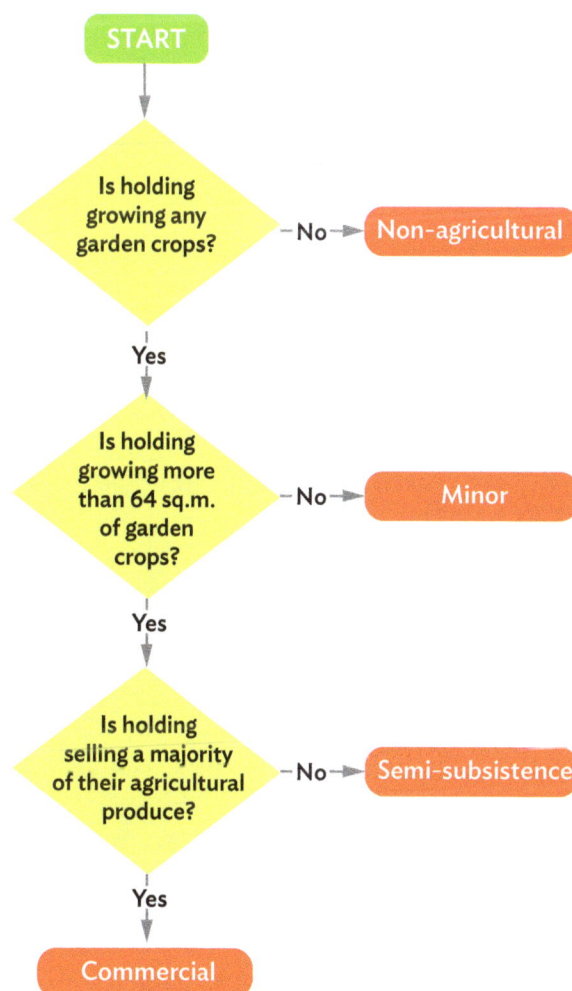

Source: Asian Development Bank visualization based on Cook Islands Census of Agriculture 2021 questionnaire.

The coefficient of variation values in the estimation of agricultural households categorized by their level of activity indicates the practicality of a sample census approach, such as increasing the number of sample enumeration areas from 12 to 20. These insights could also guide the development of a rolling census of agriculture approach in Rarotonga Island, especially given the limited personnel available to the MOA for such tasks.

Table 2: Comparison Between the 2011 Census of Agriculture and 2022 ADB Post Enumeration Survey Number of Agricultural Households by Level of Agricultural Activity Based on Census of Agriculture Definitions

Level of Agricultural Activity COA Definition	2011 COA	2022 ADB Post Enumeration Survey			
		Expansion Estimator		Ratio Estimator[a]	
		Estimate	CV (%)	Estimate	CV (%)
Agricultural HHs	1,242	1,624	23.6	1,368	13.1
Minor Agricultural	659	1,470	23.8	1,239	13.2
Subsistence Only	333	70	39.4	59	36.7
Commercial	240	83	35.2	70	28.9
Non-Agricultural	1,833	1,090	17.9	918	19.5

ADB = Asian Development Bank, COA = Census of Agriculture, CV = coefficient of variation, HH = household.

[a] Number of households engaged in agriculture as per 2021 Census of Population and Dwellings used as an auxiliary variable.

Source: Asian Development Bank estimates.

Area Measurement Methods Used in the 2022 Post Enumeration Survey Cook Islands

In this study, the unit of measurement and comparison is the agricultural parcel, which is defined as a "piece of land of one land tenure entirely surrounded by other land, water, road, forest or other features not forming part of the holding or forming part of the holding under a different land tenure type" (FAO, 2015). The agricultural parcel is selected as the unit of measurement for a few reasons. Agricultural parcel areas generally present greater consistency over time than the total area of a holding or individual crop plot level. They are less susceptible to fluctuations that might arise due to seasonal variations. This relative stability makes parcel area measurement more reliable and easier to track longitudinally. The concept of a parcel is, by definition, linked to land tenure. This connection allows for the possibility—where data are available—of cross-validating measurements with land registry data. Such cross-validation can enhance the accuracy, providing an external dataset for comparison and verification of the parcel measurements obtained in the study, if available.

Method 1: Walking the Perimeter of the Parcel with Handheld GPS

In the Cook Islands, two GPS-based measurement methods were used in this study. Field protocols used for the data collection are provided in Appendix 4: Protocols for GPS Parcel Area Measurement. The first method utilized for area measurement is a Garmin eTrex 32x, a dedicated handheld GPS device, primarily chosen for its precision and accuracy. The Garmin eTrex 32x has a rated margin of error of approximately 3.65 meters (m), indicating that the actual location from where the physical point was captured falls within a 3.65 m radius of the position recorded by the device.[1] It is important to note that this level of accuracy is expected under optimal conditions; environmental factors such as obstructions can reduce accuracy.

---

[1] Garmin eTrex 32x Owners manual.

The margin of error for handheld GPS devices is critical to monitor when measuring area. The GPS device calculates the area by tracking a series of point locations at regular intervals as the enumerator walks the perimeter of the parcel to plot its shape in GPS coordinates. A substantial margin of error at each point, especially in areas with complex boundaries, could lead to overestimation or underestimation of the actual area. Figure 4 highlights this issue, depicting how the top-right point's margin of error—being larger than the others—affects the buffer area around the true boundary of the parcel. This issue may be further exacerbated when using GPS devices with a higher margin of error.

The margin of error can also be affected by environmental factors such as rainy or cloudy weather and physical obstructions like trees and buildings. In the 2022 PES Cook Islands, approximately 12% of parcels were measured in cloudy or rainy conditions, while under 8% were measured with dense tree coverage. In cases of egregious GPS measurements, field staff often took the initiative to revisit and remeasure parcels under better conditions, and if any errors persisted were corrected during data processing.

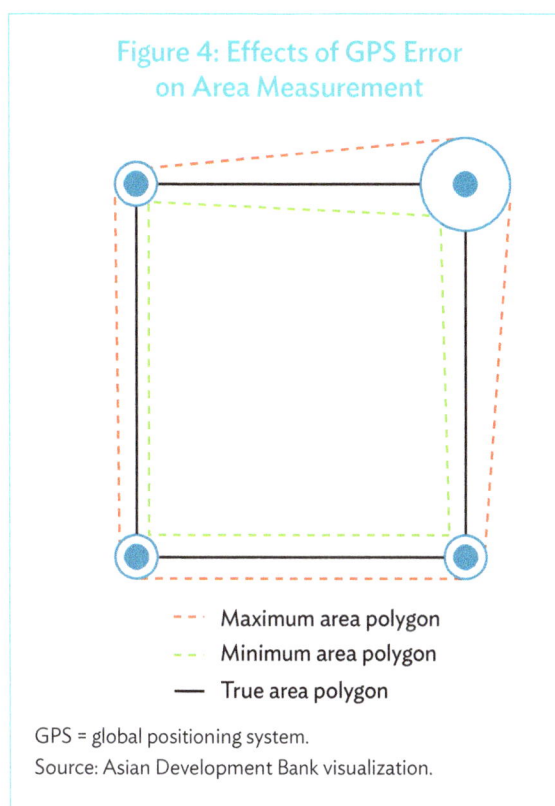

Figure 4: Effects of GPS Error on Area Measurement

– – – Maximum area polygon
– – – Minimum area polygon
——— True area polygon

GPS = global positioning system.
Source: Asian Development Bank visualization.

GPS errors are common due to sensor sensitivity to environmental conditions. Once GPS data is imported into a GIS software like Quantum Geographic Information System (QGIS), it is standard practice for a GIS analyst to review the topographical integrity of the recorded boundary data to verify that there are no overlaps or intersections. The analyst then makes necessary boundary adjustments and computes the final area in standard units in line with the local coordinate projection system. General procedures for processing GPS data are outlined in Appendix 5: Procedures for Editing GPS Data Collected for Area Measurement.

Figure 5 presents a sample of raw GPS data collected during an area measurement exercise in the Cook Islands, showcasing instances of GPS errors. These errors range from procedural oversights—such as the enumerator neglecting to end the measurement upon leaving the holding—to technical issues like backtracking along the boundary or self-intersections within the shape of the parcel caused by weak GPS signals.

GPS area measurements using consumer-grade equipment such as those used in this study are not substitutes for legally binding land cadastre or records. They serve solely for statistical comparisons against farmer-reported estimates, providing a baseline. The consumer-grade equipment used is different from commercial survey tools in accuracy, being adequate for estimating agricultural land use but not for delineating boundaries required for legal ownership.

The complexity of fieldwork and data management is an often-overlooked limitation in GPS data collection with a dedicated handheld GPS unit. As this process involves using a separate, dedicated handheld GPS device, additional training for enumerators and supervisors is necessary to operate the device proficiently. Further, the data collected—specifically the parcel boundary information—is saved directly onto the handheld GPS as a GPS Exchange Format file (GPX) file. This setup requires a manual file transfer to move the data back to the main office at the end of data collection, demanding specific field procedures and strict file naming conventions. These procedures are necessary to ensure file uniqueness and to facilitate matching with the survey microdata set. In

Figure 5: Cook Islands—Parcel Areas Captured by Handheld GPS, Garmin eTrex 32x

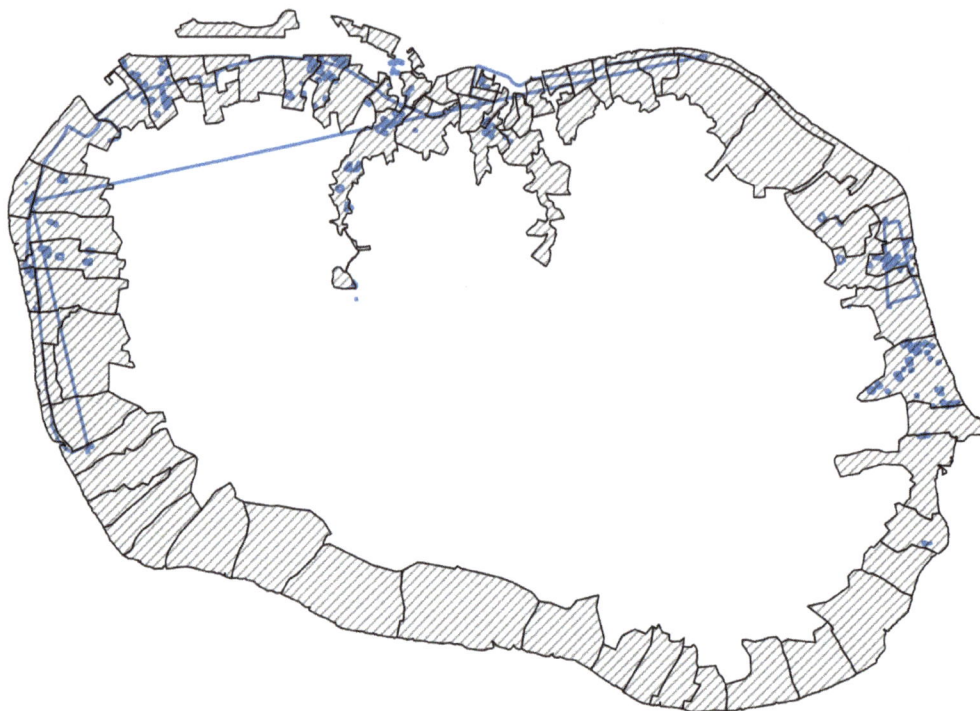

Source: Asian Development Bank visualization using Quantum Geographic Information System (QGIS) generated, August 2023.

the methodology applied in this study, a standardized naming convention based on unique household and parcel IDs was implemented to guarantee accurate record matching and linkage (Appendix 3: 2022 Cook Islands Post Enumeration Survey Questionnaire).

Additionally, enumerators were required to input the calculated area values from the GPS device into the computer-assisted personal interviewing (CAPI) application to enable real-time monitoring. This step is critical as it ensures that even if file management processes encounter issues, the area data from the GPS is accurately recorded alongside the corresponding records in the survey micro dataset. This dual approach of manual data transfer and direct data entry into the CAPI system helps maintain data integrity and accuracy in GPS-based data collection.

Despite its limitations, the use of a dedicated GPS is still considered to be a more efficient alternative to the traditional tape and compass method. The time required for measurement is based on travel time to the parcel, the availability of the farmer to guide the enumerator around the parcel, the time for checking the boundary, and walking around the parcel area. For field operations efficiency, it was also decided that parcels more than 30 minutes (by motor transport) would not be included in the measurement, which optimizes the time required for the overall data collection.

Method 2: Digitizing Parcel Boundary on a Satellite Image

In Method 2, an agricultural parcel boundary is digitized on a satellite image using the geometry (polygon manual) function in Survey Solutions version 22.12 (December 2022). An illustration of the geometry feature is shown in Figure 5. This method involves the farmer tracing the boundary of their parcel directly over a satellite image, negating the need for the farmer and enumerator to walk the boundary physically. Key to the success of this approach is the ability of the farmer to accurately recognize their land from an aerial perspective and the assumption that the satellite imagery is up-to-date and reflects the current agricultural season.

While effective for mapping agricultural parcels, the digitization method has notable limitations. Preparing the base maps for this process is time-consuming. These base maps need to be high resolution to ensure that parcel boundaries are distinctly visible, which is especially crucial for smaller parcels where finer details are necessary. The resolution requirement increases with the decrease in average parcel size, demanding more detailed and, thus, larger base map files. Moreover, the file size of the satellite images used in the base maps can become a challenge. For enumerators covering extensive areas, these large files can significantly slow down the performance of the tablets used in the field (Figure 7).

Figure 6: Digitizing an Agricultural Parcel Boundary in Survey Solutions

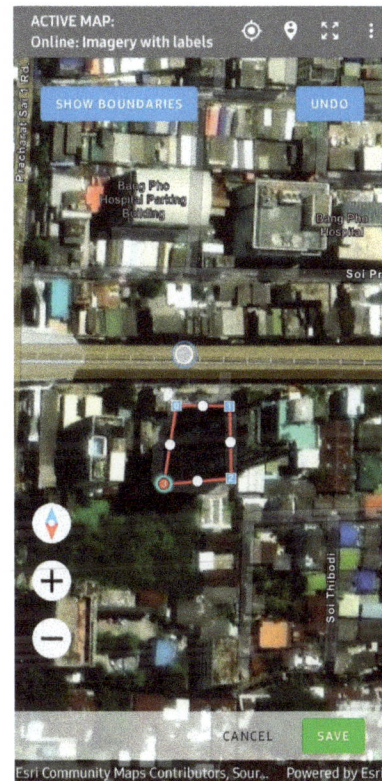

Source: Asian Development Bank visualization using Survey Solutions.

Figure 7: Cook Islands—Parcel Areas Captured by Digitization on Satellite Image Method

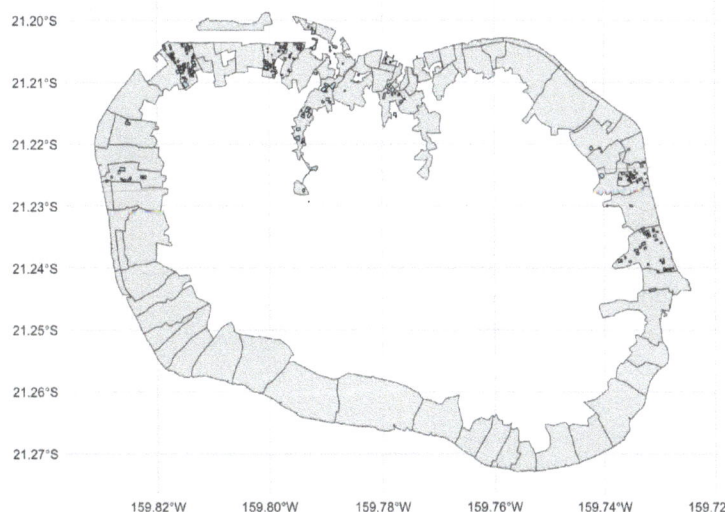

Source: Asian Development Bank visualization using data from 2022 Post Enumeration Survey Cook Islands, August 2023.

Area Measurement Methods Used in Armenia and the Lao People's Democratic Republic

Method 3: Walking the Perimeter of the Parcel with a Tablet-Based GPS Sensor

In Armenia and the Lao PDR, the approach to measuring agricultural land areas involved using onboard GPS sensors in tablets, a change from the handheld GPS devices (Garmin eTrex 32x) used in the Cook Islands. This change was facilitated by the geometry (polygon–automatic) question feature of the Survey Solution application. Outside of the handheld GPS device, the protocol for measurement remained the same as those used in the Cook Islands (Appendix 4: Protocols for GPS Parcel Area Measurement). The aim was to determine if using tablets alone could accurately and objectively measure agricultural land, which could reduce the need for extra equipment and simplify enumerator training.

Tablet GPS sensors typically offer less precision than dedicated handheld GPS devices. According to Google Android documentation, horizontal accuracy (or margin of error) is defined as a radius of 68% confidence, meaning there is a 68% probability that the actual location is within the accuracy radius from the latitude and longitude center. Therefore, in deciding to use this method, it is vital to weigh the practical benefits against the possible compromise in accuracy, ensuring that the tablets' GPS sensors are adequate for the specific demands of agricultural area measurement (Figure 8).

The error in GPS accuracy—particularly tablets—can significantly impact the measurement accuracy of agricultural parcels, especially those with irregular shapes or smaller sizes. This error in GPS readings may result in substantial discrepancies between the actual and measured area of the parcel. The issue is further exacerbated in the case of smaller parcels, where the margin of error in GPS measurements could result in a larger portion of the total area. Such a margin of error could lead to either an overestimation or underestimation of the land size. This is particularly critical in smaller parcels, where even minor inaccuracies can have a larger impact on assessing the size of the parcel. A general practice is to only require measurement of agricultural parcels larger than the margin of error of the GPS device used. For example, if the margin of error on the GPS

Figure 8. Digitizing an Agricultural Parcel Boundary by Walking Around the Parcel in Survey Solutions

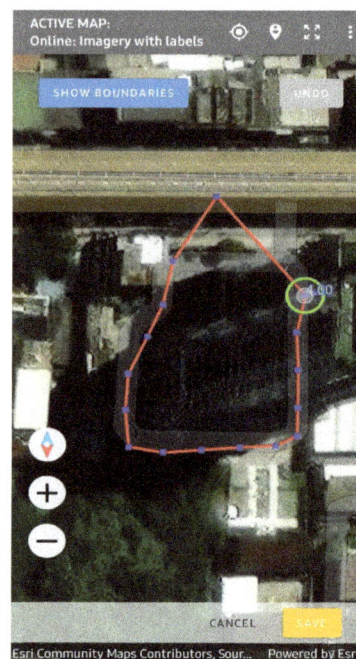

Source: Asian Development Bank visualization using Survey Solutions.

Figure 9: GPS Measurement Errors in Irregular Shapes

Source: Asian Development Bank visualization, Cook Islands Rarotonga, 2023.

sensor is 5 m, then a parcel less than 5 m by 5 m would not be measured by GPS and would be recommended to take the estimate provided by the farmer.

GPS errors become particularly notable in cases involving the traversal of tight corners or when mapping areas where two sides of the polygon are closer together than the device's GPS margin of error. This issue is demonstrated in Figure 9, illustrating how a lack of GPS accuracy can lead to difficulty mapping highly detailed corners. In the figure, the true boundary of the parcel is depicted by a red line, while the surrounding buffer indicates the error range of the GPS device used. If the GPS device has a high inaccuracy, it can lead to an inability to properly capture certain characteristics of the parcel's shape. The same problem arises in parcels featuring close parallel lines. These situations often require corrections during data cleaning, as the GPS device may not accurately capture the precise layout of such intricate boundaries.

Method 4: Digitizing Parcel Boundary on a Satellite Image by Selecting Only Parcel Corners

Method 4—an adaptation of Method 2—was designed to address the difficulties some enumerators encountered in accurately tracing the full extent of a parcel boundary. In this method, enumerators were instructed to identify and mark only the corners of the parcel using the Survey Solutions geometry multi-point question type. The primary goal of this approach was to speed up the data input process by simplifying the task to just marking key corner points. This change was intended to minimize potential errors and streamline the process, allowing enumerators to capture the essential boundaries of the parcel more easily and accurately (Figure 10).

In practice, selecting only the parcel corners proved more cumbersome than anticipated. The absence of visual indicators on the screen, such as lines connecting the points to delineate the parcel's extent, required enumerators to mentally "connect the dots." This often slowed down the data input process, as they had to visualize the parcel boundaries without on-screen assistance. A GIS analyst had to connect these corners to form the complete parcel boundary after data collection. This added step introduced additional processing time and presented a challenge in monitoring data quality in real-time.

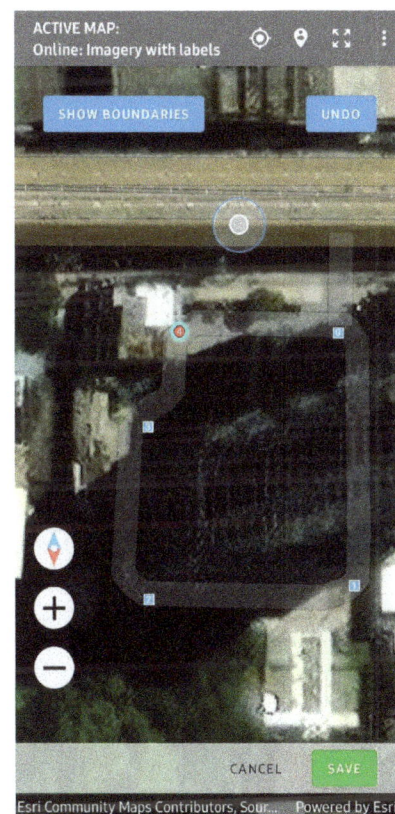

Figure 10: Digitization of Agricultural Parcel on Satellite Image by Selection of Parcel Corners

Source: Asian Development Bank visualization using Survey Solutions.

ANALYSIS OF REPORTED AND MEASURED AREA

In the following section, the study will compare the area measurements obtained through GPS with those reported by farmers for case studies in Armenia, the Cook Islands, and the Lao PDR. In cases where multiple GPS measurement methods were employed, each method will be individually compared against the farmer-reported area estimates. This comparison will be conducted at the parcel level, the primary measurement unit. The focus on individual parcels allows assessment of potential biases farmers may have when reporting the size of a single piece of agricultural land rather than considering the total aggregate area of their holdings. This approach is informed by previous studies suggesting that the degree of reporting bias is often related to the parcel size.[2] For holdings comprising both small and large parcels, the overall bias across all parcels may obscure that the bias varies depending on the size of each parcel.

Cook Islands

Estimates of total agricultural area

The 2022 Post Enumeration Survey (PES) collected farmer-reported agricultural land area data at two hierarchical levels: the overall holding and individual parcels. In the design of the questionnaire, it was expected that the aggregate of parcel areas should correspond to the total area of the holding as reported. However, several inconsistencies were reported between the two areas. In reconciling these figures during data processing, the following rules were taken: When the sum of the parcel areas surpassed the reported holding area, this sum was adopted as the corrected or reconciled holding area. Otherwise, the initially reported holding area figure was maintained. Table 3 presents these estimates in both square meters (m^2) and acres for agricultural holdings.

Table 3: Estimated Total Area in Square Meters and Acres of Agricultural Holdings

Variable Used Area Estimation	Estimated Total Area (m^2)			Estimated Total Area (acres)		
	Estimate	SE	CV (%)	Estimate	SE	CV (%)
Based on Reported Holding Area	3,377,943	639,936	18.9	834.7	158.1	18.9
Based on Reported Parcel Area	3,218,007	635,332	19.7	795.2	157.0	19.7
Based on Reconciled Reported Area	3,501,806	655,302	18.7	865.3	161.9	18.9

CV = coefficient of variation, m^2 = square meter, SE = standard error.

Source: Asian Development Bank estimates based on 2022 PES Cook Islands.

---

[2] Bias is defined as abs (Farmer-reported area: GPS measured area)/ GPS measured area * 100.

Comparatively, the total area of parcels recorded in the 2011 agriculture census was 569.7 acres. The reconciled figure of 865.3 acres—derived from the 2022 PES—indicates an expansion in agricultural land area on the island of Rarotonga over the previous 11 years. Such an expansion would be consistent with the growth of the agriculture sector over the same period.

Table 4 shows area estimates for all holdings—comparing traditional farmer-reported and GPS methods—and focuses on matched observations (parcels with both reported and GPS-measured areas). That is, only parcels with both reported and GPS-measured areas were considered to generate estimates. Unfortunately, the GPS area measurement was not conducted in some parcels for various reasons such as being very far from the responding holding. Weight adjustments were applied to these estimates to account for missing measurements.

The estimated total area of agricultural holdings using the GPS-measured area was slightly higher than the farmer-reported method of area measurement. Table 4 reveals a general trend: GPS-measured areas yield higher land area estimates than farmer reports across all farmer categories: commercial, minor agricultural, and subsistence. Specifically, GPS readings estimated the total area at 858.9 acres with a standard error of 172.3 and a coefficient of variation of 20.1%, higher than the 769.7 acres estimated through farmer reports. This discrepancy suggests underreporting in traditional farmer-reported methods.

Table 4. Comparison of Estimated Total Area of Holdings by Reported and GPS-Assisted Measurements

Methodology Used for Area Measurement of Holding	Estimated Total Area (m²)			Estimated Total Area (acres)		
	Estimate	SE	CV (%)	Estimate	SE	CV (%)
Based on Reported Area	3,115,017	568,551	18.3	769.7	140.5	18.3
Based on GPS Readings	3,475,991	697,194	20.1	858.9	172.3	20.1
Commercial Growers						
Based on Reported Area	741,911	332,303	44.8	183.3	82.1	44.8
Based on GPS Readings	922,156	413,986	44.9	227.9	102.3	44.9
Minor Agricultural						
Based on Reported Area	2,217,129	379,564	17.1	547.9	93.8	17.1
Based on GPS Readings	2,394,520	429,708	17.9	591.7	106.2	17.9
Subsistence Only						
Based on Reported Area	155,977	85,648	54.9	38.5	21.2	54.9
Based on GPS Readings	159,315	88,942	55.8	39.4	22.0	55.8

CV = coefficient of variation, GPS = global positioning system, m² = square meter, SE = standard error.
Note: Based on matched records. That is only for holdings and parcels with both reported and GPS-measured areas. Weight adjustments were applied for the matched data file.
Source: Asian Development Bank estimates based on 2022 Post Enumeration Survey Cook Islands.

Comparisons of farmer-reported versus GPS-measured areas at parcel level

Farmer-reported parcel areas in the Cook Islands were notably smaller than those obtained through GPS measurements, with an average discrepancy of 184.56 m². This underreporting is statistically significant (p = 0.02) and falls within a 95% confidence interval of −340.7 m² to 28.42 m². Given that the average parcel size in the Cook Islands is approximately 0.35 acres or 1,416 m², the observed mean difference of −184.56 m² represents an underestimation of about 13% of the average parcel size.

Figure 11 depicts a scatter plot, which displays the farmer-reported areas on the x-axis and the GPS-measured areas on the y-axis. The black diagonal line represents the 45-degree line where the reported and measured areas are equal. Visually, the data points suggest an inconsistency in the level of underreporting or overreporting across different parcel sizes. Parcels smaller than 2,000 m² are often reported as larger than the GPS measurements suggest, indicating overreporting. Conversely, parcels larger than 2,000 m² tend to be underreported. The study proposes quantile regression analysis to examine this pattern across the range of parcel sizes. This technique will allow for a more detailed investigation into how the discrepancies between reported and measured areas vary at different points of the parcel size distribution.

Figure 11: Cook Islands—Farmer-Reported and Digitized Areas Versus GPS Area Measurement

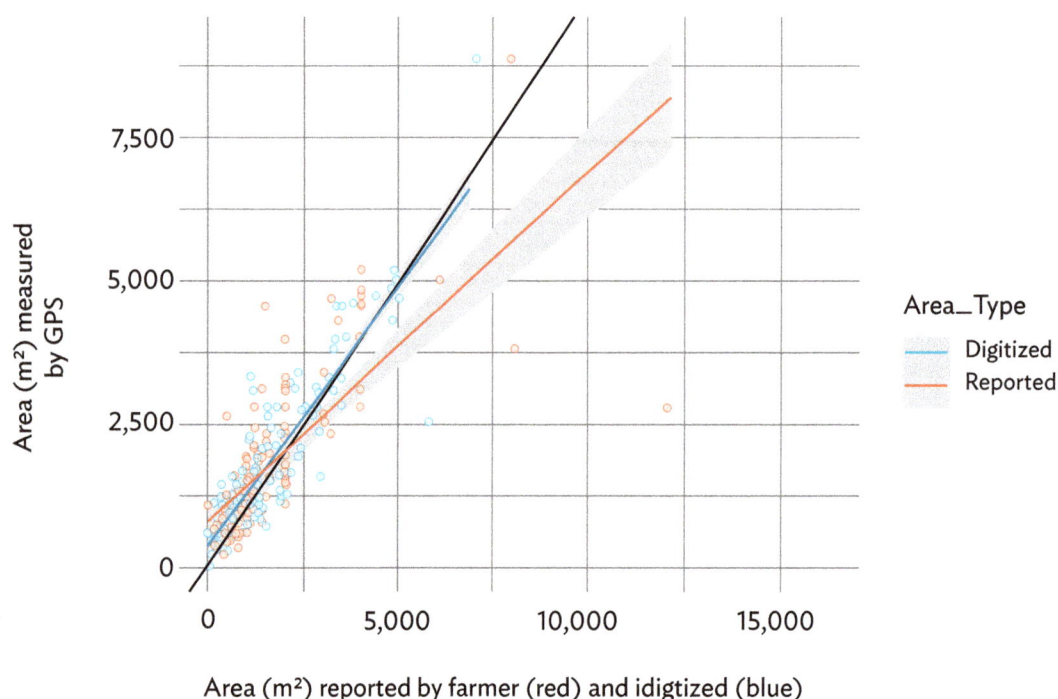

GPS = global positioning system, m² = square meter.
Source: Asian Development Bank estimates based on 2022 Post Enumeration Survey Cook Islands.

Quantile regression offers an approach to analyze the relationship between farmer-reported and GPS-derived parcel measurements across various segments of the distribution, specifically at the 10th, 25th, 50th (median), 75th, and 90th percentiles. By employing this method, the study aims to discern if the discrepancies between farmer-reported sizes and those measured by GPS are uniform across the entire range of parcel sizes or if they are more pronounced at certain points in the distribution. For instance, larger discrepancies might be identified at the lower or upper tails of the distribution, which could suggest different levels of reporting bias or measurement error in parcels of different sizes. The coefficients of this quantile regression approach are presented in Figure 12.

At the 10th percentile—representing the smallest parcels in the sample—there is a statistically significant positive relationship between the two measurements: for every one-unit increase in farmer-reported measurements there is a 0.436-unit increase in the GPS-derived measurements. This is statistically significant, as the 95% confidence interval ranges from 0.0776 to 0.7944. This suggests that farmers may overestimate their parcel sizes compared to the GPS-derived measurements, particularly at the lower end of the parcel size distribution.

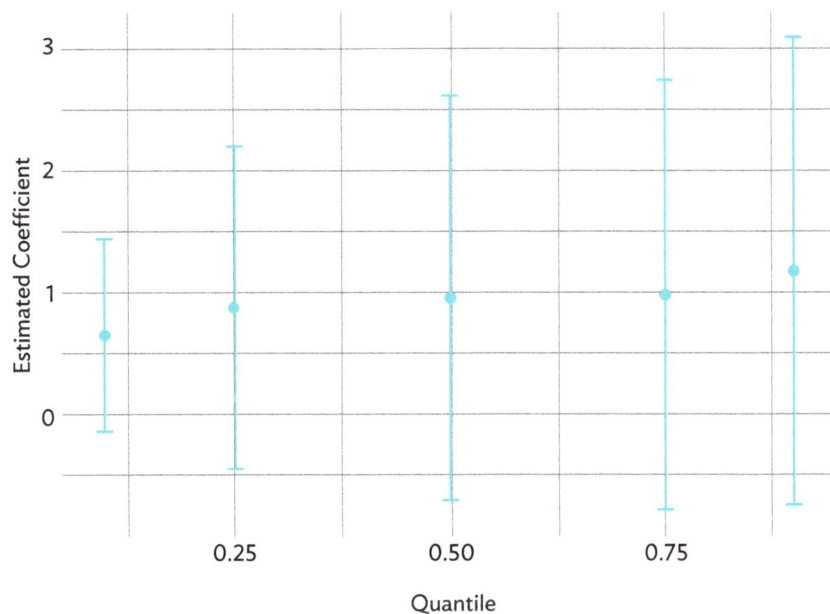

Figure 12: Quantile Regression Coefficients Comparing Farmer-Reported and GPS Area Measurement

GPS = global positioning system.
Source: Asian Development Bank estimates based on 2022 Post Enumeration Survey Cook Islands.

Moving to larger parcels, this relationship becomes less certain. At the 25th, 50th (median), 75th, and 90th percentiles, the estimated coefficients suggest a positive relationship, indicating that larger farmer-reported measurements are associated with larger increases in GPS-derived measurements. Yet, the correlations are not statistically significant at these quantiles.

Although the aggregated data suggests a general underestimation of parcel sizes by farmers, the dynamics of this bias vary significantly when parcel size is considered. Specifically, at the 10th percentile, farmers overestimate parcel sizes, indicated by a ratio of 1:0.436. Larger parcels may predominantly drive the underestimation observed in the initial analysis across all parcels. Noting the lack of statistical significance in estimated coefficients for larger parcels, it may be considered a recommendation for the future, where the sample should include all or a higher allocation of larger commercial farms to study this relationship further.

Conversely, when comparing farmer-reported data to digitized methods, the reported measurements were, on average, 155.17 units higher. This difference also shows statistical significance, with a p-value of 0.03 and a 95% confidence interval between 12.37 and 297.97, highlighting that farmers report larger areas than captured by the digitization method.

Digitized methods for measuring land areas have advantages over using GPS to traverse the perimeter of the parcel. In the Cook Islands, these methods resulted in collecting a higher number of measured observations. Reasons for undertaking the GPS method include instances where the farmer could not join the enumerator at the parcel, situations where walking the parcel was unsafe, and cases where farmers were reluctant to allow enumerators on their land, fearing crop damage.

However, these methods need less error correction because they do not have the same issues as GPS sensors. However, even if digitized maps look neat, they might not always indicate accuracy. Sometimes the enumerators outlining the parcel boundaries on satellite imagery might wrongly follow the visual boundaries seen on the map instead of the actual field markers and boundaries.

Armenia

In Armenia, the digitization method for land measurement posed distinct challenges. A primary issue was the farmers' difficulty in identifying their parcels on satellite images, as highlighted in Figure 13. This figure shows two parcels: one outlined by physically walking the parcel (green) and the other through digitization on a satellite image (red). The green, walked-around parcels usually match the actual parcel shape more accurately. In contrast, the digitized (red) parcels often differ in scale (either larger or smaller) and may be offset from their actual location. Certain obstructions—including the presence of water puddles, fallen trees, and inaccessibility due to fencing—were observed by the enumerators when measuring the land parcel by walking method.

Figure 13: Differences in the Perception of Scale as Captured by Walking (Green) and Digitization (Red) Methods

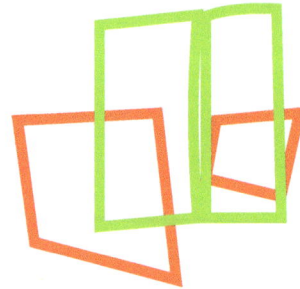

Source: Asian Development Bank visualization using Quantum Geographic Information System (QGIS).

This mismatch could partly be due to inadequate enumerator training in Armenia of 1 day in a classroom setting, unlike in the Cook Islands, where enumerators spent up to 1 week familiarizing themselves with the tablet and its area measurement functions. Additionally, the lack of detail on the base map contributed to these issues. Due to internet constraints, enumerators had to work with slightly blurry satellite imagery. The base map—derived from Sentinel-2 data with a 10 m spatial resolution—was sufficient for locating major features like dwellings, farm buildings, and roads. However, it was less effective in delineating precise parcel boundaries in smaller land parcels. Consequently, enumerators sometimes switched to an online map from the Survey Solutions platform, which was clearer but might not have accurately reflected current seasonal conditions.

Though limited in scope to Berdik village, the pilot study in Armenia offers useful insights into the use of GPS measurements in an area with a well-defined land cadastral system. Figure 14—resembling the scatter plot from the Cook Islands study—reveals an interesting observation: despite some differences, the area reported by farmers is closely aligned with the GPS measurements using both the walking and digitized methods. This consistency suggests that the established land cadastral system of the village might have given farmers a reliable land size baseline. However, due to the age of this cadastral system, some land parcels were observed to have been divided or combined over time. In these instances, the farmers-reported land sizes showed variance from the GPS-measured areas.

It is also important to note the sensitivities surrounding GPS data collection in the broader context across Armenia, the Cook Islands, and the Lao PDR. Farmers may not always welcome activities that involve walking across their land, potentially damaging crops. Enumerators must be considerate, avoid property damage, and respect privacy rights. There was apprehension about sharing geolocation data in Armenia due to conflict-related concerns. Farmers hesitated to provide GPS coordinates of their homes and properties, fearing they could be targeted. Moreover, some of the land parcels where farmers were practicing subsistence farming and growing crops for household consumption had their private residences as well, which was one of the reasons for farmer reluctance to area measurement. This highlights the need for enumerators to approach such tasks with sensitivity and respect

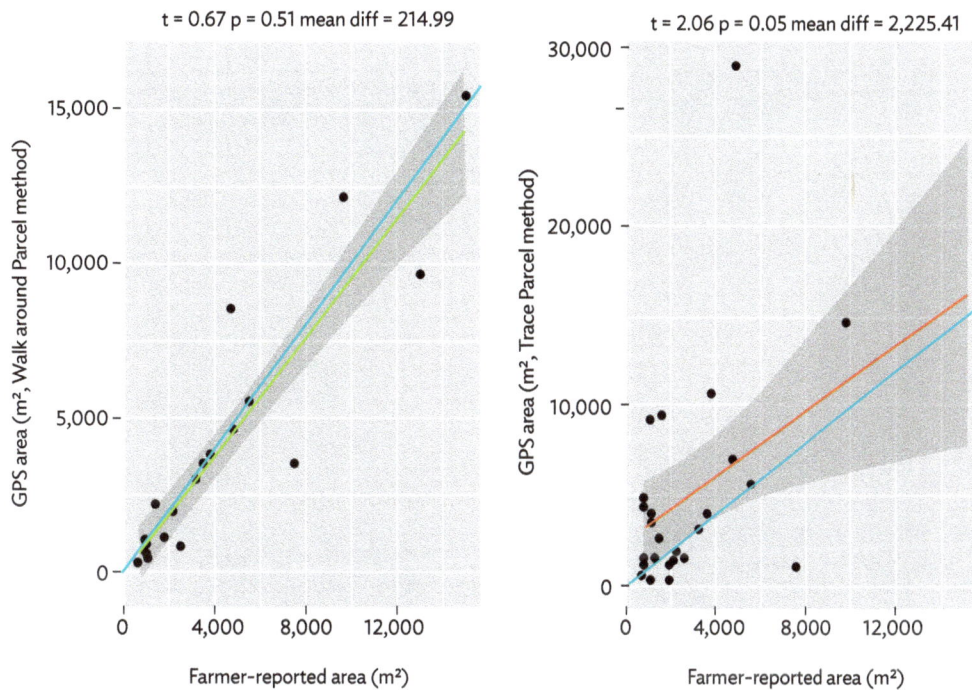

Figure 14: Armenia Farmer-Reported and Digitized Area Versus GPS Area Measurement

t = 0.67 p = 0.51 mean diff = 214.99

t = 2.06 p = 0.05 mean diff = 2,225.41

GPS = global positioning system, mean diff = difference of means, m² = square meter, p = probability, t = Student's t.
Source: Asian Development Bank estimates based on 2023 area measurement pilot in Armenia.

for local contexts and concerns. It also points to the sensitive nature of geospatial data and the need to ensure the confidentiality and privacy of any location-based information collected.

Lao PDR

In the Lao PDR, three methods for measuring land area were tested: walking around the parcel, digitizing parcel boundaries on satellite imagery, and selecting point corners. The results from the pilot area of Pak Pok village are presented in Figure 14.

During the GPS area measurement method—which involved walking around the parcel—enumerators faced several challenges with their devices. Issues arose from the quality of GPS sensors and the difficulty in handling large satellite images, especially on slower devices. These problems could be attributed to the variety of tablet devices used in the field. The range included both newer and older models, as well as personal devices brought by the enumerators themselves. To address these inconsistencies and enhance the uniformity of data collection, it was recommended following the pilot to standardize the devices used by all enumerators. Having a consistent set of devices would help ensure uniform accuracy and precision in the GPS measurements used.

Some of the challenges faced during area measurement by walking method were: obstruction of parcel corners due to thick vegetation that the enumerators could not walk around, the presence of water in the fields—which made certain corners of land plots inaccessible—and that certain sections of the parcels did not have any room for the enumerators to walk on and take measurements.

Enumerators favored the digitization method for its ease of collection but faced difficulties in areas with poor internet connectivity. Preparing for the survey demands extra time to select and ready offline maps for all areas sampled.

Figure 15: GPS Parcel Areas Captured in Pak Pok Village, Vang Vieng, Vientiane Province, Lao People's Democratic Republic

GPS = global positioning system, ESRI = Environmental Systems Research Institute.
Source: Asian Development Bank visualization using Quantum Geographic Information System (QGIS).

Effective pre-planning is imperative for using offline maps. This step would require accurately mapping out the specific coverage area of each enumerator and ensuring the offline map is uploaded to the device before data collection. This step must also ensure the map loads properly on tablets without technical issues.

Public satellite imagery sources—such as Google Earth or Environmental Systems Research Institute, Inc.—were also noted to have limitations, particularly when the date of the imagery does not align with the data collection period. To mitigate this, it is advisable to use satellite images closer to the data collection season, or high-resolution data from government sources—like UAV imagery—could be prepared.

Analyzing the data from the walking method compared to farmer-reported areas, it is evident that there is a wide variation in the discrepancies between reported and GPS-measured areas, suggesting an inconsistency in the accuracy of farmers' reporting (Figure 16). Some farmers significantly underestimated their land size, with biases as large as −93.2%, while others overestimated, with positive biases up to 37.4%.

A trend of underestimation by farmers is apparent despite the limited sample size. In numerous instances, the area reported by farmers is considerably lower than that measured by GPS. Reasons for this could include a lack of precise knowledge about the size of their land or intentionally conservative estimates, possibly influenced by concerns over higher taxation by the government on larger land sizes.

Figure 16: Lao People's Democratic Republic Farmer-Reported and Digitized Area Versus GPS Area Measurement

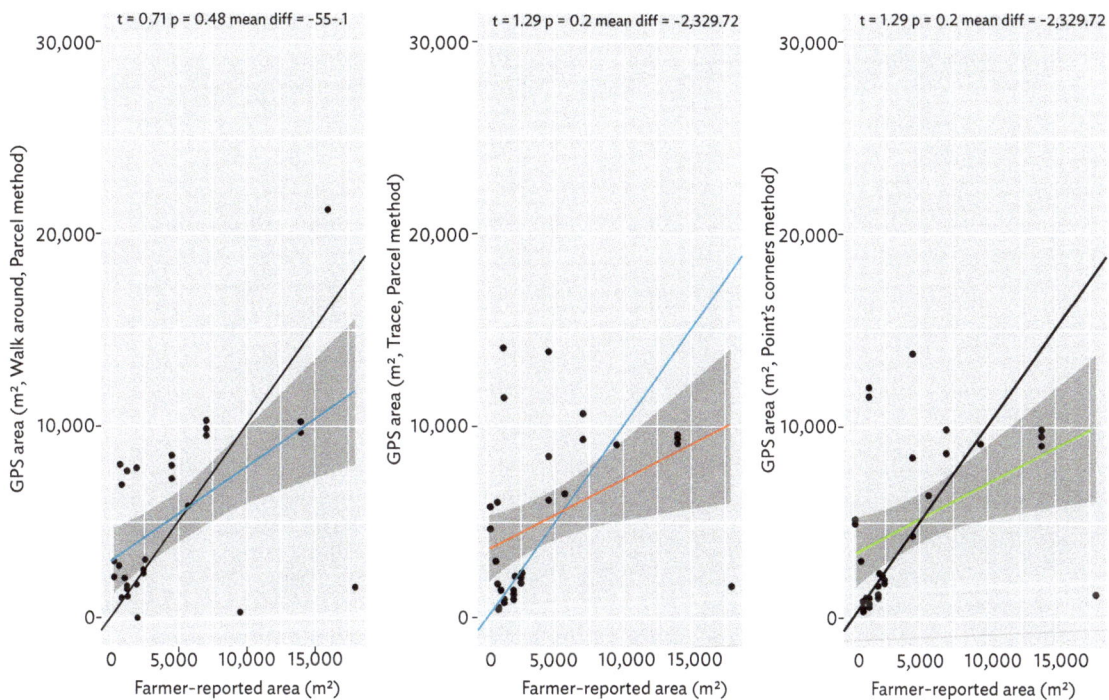

GPS = geographic positioning system, mean diff = difference of means, m² = square meter, p = probability, t = Student's T.

Source: Asian Development Bank estimates based on 2023 area measurement pilot in the Lao PDR.

The data also shows that larger parcels often have more significant discrepancies between reported and measured sizes. For instance, parcels reported as 14,000 m² displayed a bias of 37.4% and 16,000 m² displayed a bias of −24.7%. This suggests increasing difficulty in accurately estimating the size of larger parcels, potentially due to their complexity or the challenges of visually assessing larger areas.

Discussion

When considering the most suitable method for GPS land measurement, several critical factors—such as the size, shape, and terrain of the parcel—must be considered, along with the resources at hand. While dedicated handheld GPS devices are often regarded as the gold standard due to their high precision and reliability, they come with data management and processing challenges. Despite these challenges, their use is still recommended when the highest positional accuracy and measurement precision are required. Nonetheless, there are cases where utilizing GPS measurement methods integrated into tablets can be advantageous, particularly considering their convenience and potential for integration with other data collection tools.

The walking method—whether using a dedicated handheld GPS device or on-tablet GPS sensor, where enumerators physically traverse the perimeter of the parcel—is particularly effective for smaller parcels with complex shapes and easily navigable terrain. This method allows for precise boundary capture but can be time-consuming for larger parcels, potentially taking up to 1 hour for areas exceeding 10,000 m².

Conversely, the digitization method is more suitable for large, monocropped areas. These parcels are often well-defined and easier to identify on a map, making digitization a time-efficient alternative to walking, especially in regions with access to accurate and high-resolution satellite imagery. However, the effectiveness of this method is dependent on the quality of the base maps and satellite data.

The parcel corner GPS method—which involves marking only the parcel corners—is generally not recommended based on the experiences in this study due to various field challenges, such as difficulties in accurately identifying corner points, particularly in irregularly shaped parcels. This can lead to significant inaccuracies in area measurement if the enumerators are not properly trained and well-versed in using the field instrument.

Scaling GPS area measurement methodologies nationally requires focusing on several factors: building geospatial capacity in national statistics offices and agricultural line ministries; sufficiently training field staff in GPS technology usage; and managing logistical challenges of data management across the tablet where information is recorded, and the GPS device where the area boundary files. As demonstrated in the pilots of this study, it is crucial to ensure that data are effectively transferred, reviewed, and consolidated at all levels.

After collecting these data, developing a knowledge management plan is essential for utilizing this data in agricultural statistics. Such an application would typically involve deriving adjustment factors—ratios of GPS-measured areas to farmer-reported estimates—to account for the bias resulting from the farmer-reported estimates. These factors can be tailored by parcel size, region, or agricultural practice to reflect accurately variations throughout a country. These revised adjustments, in turn, could support policymakers in improving resource allocations through extension services to farmers and better plans to meet food security requirements.

IMPLICATIONS FOR POLICY AND RECOMMENDATIONS

Enhancing objective agricultural area measurement practices requires a comprehensive approach. A primary recommendation is the standardization of objective measurement practices and integration in agricultural surveys where land is a key indicator. Depending on the reliability of the farmer-reported data, it is recommended that national statistical offices or line ministries involved in collecting agricultural statistics conduct an objective measure survey to compile a set of conversion or adjustment factors related to the biases in farmer-reported data. These adjustment factors—calculated as the ratio of GPS-measured area to farmer-reported area—could be provided in aggregate or disaggregated by parcel size or geographic area. These adjustment factors could be applied in relevant analyses and help to inform policies for improved planning.

Selecting the most appropriate measurement method—walking, digitization, or parcel corner GPS—should be carefully considered. This selection should be based on specific characteristics pertinent to the local context, such as parcel size, shape, and terrain. Such tailored approaches will contribute to more accurate and reliable data. Developing detailed, clear guidelines for each measurement method is also key, addressing diverse terrains and parcel sizes and providing strategies for challenges like low internet connectivity in remote areas. It is also feasible that more than one area measurement method could be deployed depending on the diversity of agricultural practices in the country.

Investment in technology and human resources to use them is crucial, with a need for high-quality GPS devices and access to updated satellite imagery linked with comprehensive training for enumerators and farmers to ensure accurate data collection. National statistical offices in Asia and the Pacific still lag in adopting geospatial technology and its uses for agricultural statistics. There exists an opportunity to share best practices in the use of the technology so that these technologies become a regular part of ongoing data collection activities in the region.

Encouraging ongoing research and development in land measurement methods and looking into new technologies and advanced imaging techniques is important. Technology is ever evolving. Over time, the sensors to collect geospatial information will become increasingly accurate and cost-effective. Imageries used for digitization will likewise be cheaper, high resolution, and available on a timely basis. Opportunities to investigate ultra-high resolution satellite imageries originating from UAVs or specialized satellite sensors allow for improved area measurement methodologies and specialized analyses such as crop health analysis or standing crop counts.

CONCLUSION

In conclusion, this study highlights the importance of accurate land area measurement in agricultural and land management policies. Accurate land area data is indispensable for a comprehensive understanding of the agriculture sector and for informed policymaking by ministries. By recognizing and addressing biases in these measurements, policymakers can gain a more accurate perspective on agricultural output. This, in turn, enables them to tailor extension services more effectively, both under normal conditions and in response to disasters.

Consideration of different measurement methods—walking, digitization, and parcel corner GPS—reveals that each approach has distinct advantages and limitations for implementation by agricultural data producers looking to introduce objective area measurement methods suitable for their agricultural statistical activity and budget. These methods offer a range of options for producers of agricultural statistics to consider in developing a survey using an objective method for land area measurement.

Future research and ongoing advancements in land measurement techniques—including exploring emerging technologies such as UAVs or ultra-high resolution satellite imagery—are critical for enhancing efficiency and precision. Adopting these technologies in agricultural statistics enables the implementation of area measurement methods on a larger scale, as they simplify and reduce the cost of conducting such surveys. This evolution in methodology holds promise for more accessible and accurate agricultural data collection.

2022 COOK ISLANDS POST ENUMERATION SURVEY DESIGN

Post Enumeration Survey Sample Design for the 2022 Cook Islands Census of Agriculture for Rarotonga

The Census of Agriculture (COA) for the Cook Islands was conducted in 2022 by the Ministry of Agriculture (MOA). The last COA was conducted in 2011. As in the previous COA, a post enumeration survey (PES) was conducted from September to October 2022 for the island of Rarotonga which is the major island of the country. Such operation was made possible with assistance from the Asian Development Bank (ADB) in cooperation with the Cook Islands MOA as well as support from the Cook Islands Statistics Office (CISO).

The PES aims to provide measures of survey quality of selected indicators from the 2022 COA. In particular, the PES aims to provide a measure of the extent of coverage errors in the last COA. In addition, the 2022 PES introduced the use of Global Positioning System (GPS) devices to obtain area measurements for the parcels of land reported. This should provide valuable data on the difference between a more objective manner of area measurements compared to reported areas. This could pave the way in assessing possible limitations on the use of the 2022 COA as well as testing the feasibility of the utilization of GPS devices in reporting area measurements, which is a vital indicator for the agriculture sector.

Coverage

For this PES, the survey population consists of all households Rarotonga enumerated in the 2022 Census of Population and Dwelling (COPD) and classified as households with agricultural activity, i.e., growing crops (fruit trees, vegetable crops, root crops) and/or engaged in floriculture, livestock raising, fishing activities including pearl gathering and/or farming. The reference period used is the calendar year 2021. The COPD was conducted by the CISO under the Ministry of Finance. In effect, the PES—on a sample basis—would adopt the original plan for the conduct of the COA. That is, agricultural households were identified.

By choosing this survey population, the PES intends to measure non-sampling errors that are associated with the 2022 COA that was conducted by the MOA. In particular, the non-sampling errors that the PES intends to measure are coverage and measurement (for land area). According to the original plan for the conduct of the COA contained in the enumerator's manual [2], it should follow the methodology adopted in 2011[3]. That is, the enumeration for the COPD is first conducted from each enumeration area by the CISO. During that fieldwork, the CISO asks questions about whether the household engages in some agricultural activities. Thus, the CISO would provide a list of households engaged in some kind of agricultural activity to the MOA from which the COA questionnaire is administered. From the initial list provided by the CISO to the MOA, there were 2,286 households engaged in agricultural activities as defined in the COPD for Rarotonga. Of this number, 1,847 were growing crops.

For the 2022 COA, the MOA deviated from the planned approach and instead conducted the census without the list provided by the CISO. The MOA census team produced a list of farm households. This list came from individuals who at some time approached the ministry for some form of assistance such as planting materials, fertilizers, and other inputs as well as advisory matters. Additional names were added to this list based on personal acquaintances of the staff of the ministry. Based on initial statistical tables shared by the MOA, there were 302 households identified as agricultural households for Rarotonga that were enumerated during the 2022 COA. Of these households, 261 were classified as agricultural households as per MOA definitions.

In 2011, the COPD enumerated about 2,098 households engaged in agricultural activities for Rarotonga. The list of these households served as a list frame that the MOA used during the 2011 COA. From this number, the MOA—after application of their definitions—identified 1,232 households as farm households (primarily growing crops). Based on this, it appears that there was a substantial drop in the number of farm households (as per the MOA definition) covered in the 2022 COA. Such a drop may be due to the MOA's decision to forgo the original plan as to how the COA should be conducted.

There are conceptual differences in the coverage of households engaged in agricultural activities between the COPD and the COA. In the COPD, agricultural activity is defined as growing crops (fruit trees, vegetables, root crops), poultry and livestock, fishing, and floriculture. The COA definition was primarily crops with an area threshold of at least 64 m^2 of land and/or 20 tree crops devoted to such activity. Livestock and poultry were treated separately in the COA. Some households were identified as agricultural households based on personal knowledge, which may not align with the CISO's definition of a household engaged in agricultural activities.

The COPD classifies households into four categories of agricultural activity: Subsistence Only, Commercial Only, Subsistence and Commercial Only, and No Agriculture. The COA classification of households was primarily based on crop production as Non-Agriculture, Minor Agricultural, Subsistence, and Commercial.

Based on the planned approach, the households with agricultural activity included in the COA can be considered a subset of the COPD. Thus, when the COA deviated from the original approach to the census, it opened up the possibility of non-coverage. Moreover, there is also the possibility that households covered in the COPD not classified as having agricultural activity may have such activity. However, based on the PES results in 2011, such instances are considered rare occurrences, and considering that Rarotonga can be considered a small geographic area in terms of statistical operations, few households may be missed. Therefore, for this PES, it was decided to cover all households with agricultural activities as covered in the 2022 COPD.

Proposed Sampling Design

For purposes of this PES—as well as considerations on the resources available for this undertaking—a stratified cluster sampling design is proposed. That is, the 78 enumeration areas for Rarotonga will first be grouped into 3 strata utilizing information from satellite imagery processed by the team utilizing initial satellite imagery from the European Space Agency [4]. From the initial European Space Agency maps provided, an estimation of the total enumeration area as well as an estimation of area potentially for agricultural use for each of the 78 enumeration areas were generated.

The resulting data extracted and/or generated forms part of the area frame. A brief description of how the area frame was generated is presented in Appendix 2 [5]. Some descriptive statistics of key variables for the design of the PES are shown in Table A1.1. Table A1.1 shows that on average, each enumeration area comprises about 30 households engaged in some agricultural activity (crop growing, livestock raising, floriculture, fishing activities). Of these, about 24 households on average are engaged in the growing of crops.

Table A1.1: Descriptive Statistics of Selected Enumeration Area Characteristics (Rarotonga, Cook Islands 2021, 78 Enumeration Areas)

Variable	Mean	Standard Deviation	CV (%)
Number of HHs engaged in Agriculture	29.31	17.36	59.24
Number of HHs growing crops	23.68	15.37	64.89
Estimated Total Area (ha) for Agriculture***	21.48	17.57	81.81
Estimated Percentage of Area for Agriculture***	70.31	16.45	23.40

CV = coefficient of variation, ha = hectare, HHs = households.

*** Estimated using satellite imagery.

Note: The count of agricultural households is derived from the 2021 Census of Population and Dwellings, Cook Islands, based on a series of questions related to the engagement in agriculture through activities in crops and livestock. Estimates for the area under agriculture are derived through a remote-sensing approach (main text, p. 14).

Source: 2021 Census of Population and Dwellings, Cook Islands.

A list frame of enumeration area characteristics derived from the 2022 COPD was prepared, and the area frame information was incorporated in such frame for purposes of sample selection and stratification. The sampling frame prepared is shown in Appendix 1. It is well known that stratification improves the precision of estimates for a fixed sample size provided that the stratification variable is correlated with major survey variables. For the PES, the important variables considered—referred to as design-test variables—are the total number of households growing crops and the total number of households engaged in agricultural activities. The stratification variables from the area frame considered are the estimated total area of enumeration area for agricultural use as well as the percentage of total enumeration area that may be utilized for agricultural activities. The correlation matrix for these variables is shown in Table A1.2. This table shows that the total enumeration area that can be potentially used for agricultural activity exhibits moderate correlation with the major variables considered and hence is chosen as stratification variable.

Table A1.2: Correlation Matrix Between Design-Test Variables (A1, A2) and Stratification Variables (A3, A4)

Variable Label	Variable	A1	A2	A3	A4
Number of HHs engaged in Agriculture per EA	A1	1.00	0.96	0.61	-0.05
Number of HHs growing crops per EA	A2	0.96	1.00	0.62	-0.01
Estimated Total Area (ha) of EA for Agriculture (Stratification Variable)	A3	0.61	0.62	1.00	0.50
Estimated Percentage of EA for Agriculture (Stratification Variable)	A4	-0.05	-0.01	0.50	1.00

EA = enumeration area, ha = hectare, HHs = households.

Source: 2021 Census of Population and Dwellings, Cook Islands and authors calculations.

To achieve close to optimal stratification (that is, determination of stratum boundaries that will achieve higher gains in precision), the cum rule by Dalenius and Hodges (1959) was used to group the 78 enumeration areas in three strata. The stratum boundaries using this rule are as follows:

(i) Stratum 1 (enumeration areas with a relatively small area [hectares (ha)] that can potentially be used for agricultural activities) All enumeration areas whose total potential area for agricultural activities is less than 10.93 ha. This stratum included 27 enumeration areas .

(ii) Stratum 2 (enumeration areas with a relatively moderate area [ha] that can potentially be used for agricultural activities) – All enumeration areas whose total potential area for agricultural activities is between 10.93 ha and 31.18 ha. This stratum included 30 enumeration areas .

(iii)   Stratum 3 (enumeration areas with a relatively larger area [ha] that can potentially be used for agricultural activities) – All enumeration areas whose total potential area for agricultural activities is greater than 31.18 ha.

These strata are designed to represent the "small," "medium," and "large" enumeration areas in terms of the potential area of agriculture as derived in Figure 2. A selection of three strata was chosen to be most optimal considering the total number of enumeration areas of 78. The stratum boundaries were derived using the cumulative square root of frequency (Cum √F), established by Dalenius and Hodges (1959), where the aim is to minimize variability within the strata and maximize variability between them. The procedure for implementing the rule is as follows:

(i)   Sort the enumeration areas based on the potential of the agricultural area (acres).

(ii)   Calculate the frequency of each unique value in the list of enumeration areas.

(iii)   Calculate the square root of the frequency for each unique value.

(iv)   Segment the list of enumeration areas such that the cumulative square root of the frequency totals for each enumeration area is equal.

Table A1.3 presents the mean and coefficient of variation per enumeration area belonging to each stratum for the design-test variables of the number of households with agricultural activities and the number of households growing crops. As expected, the average number of households with agricultural activities as well as the number of households growing crops increases with the increase in the potential area of the enumeration areas for agriculture. It can also be noted that the coefficient of variation values in each stratum are lower than the coefficient of variation values shown in Table A1.1. This indicates that a gain in precision of the estimates can be achieved with the use of the total potential area for agriculture as a stratification variable.

Table A1.3: Mean and Coefficient of Variation Values of Design-Test Variables by Strata

Stratum	Potential Area for Agriculture (ha)	Number of EAs	Number of HHs with Agricultural Activity		Number of HHs Growing Crops	
			Mean	CV (%)	Mean	CV (%)
1	< 10.93	27	16.04	38.07	12.22	43.73
2	10.93 – 31.18	30	29.20	43.19	23.73	45.72
3	>31.18	21	46.53	39.18	38.33	48.08

CV = coefficient of variation, EA = enumeration area, ha = hectare, HHs = households.
Source: 2021 Census of Population and Dwellings, Cook Islands and author's calculations.

Sample Size Determination

Of interest in the PES are the estimations of the population parameters of the survey population such as the population total. In particular, of interest for this PES would be the estimation of the total number of households engaged in agriculture as well as the total number of households growing crops as per the COA definition. In the calculation of the sample size, it is assumed that the required number of sample clusters required under simple random sampling is based on a targeted level of precision. The final choice for the sample size will also take into consideration the resources available for the PES (personnel, time, and cost).

Under a one-stage cluster sampling design the population total is defined as:

$$Y = \sum_{i=1}^{A}\sum_{j=1}^{B_i} y_{ij} = \sum_{i=1}^{A} Y_{i.} = A\overline{Y}, \quad \overline{Y} = \frac{1}{A}\sum_{i=1}^{A} Y_{i.} \quad (1)$$

Where y_{ij} is the value of the characteristic y from household j in enumeration area i; $Y_{i.} = \sum_{j=1}^{B_i} y_{ij}$ is total of the characteristic y from all households in enumeration area i; characteristic of interest (e.g. total number of households engaged in agriculture in enumeration area i) in cluster (enumeration area) i; and A are the total number of clusters. In addition, $\overline{Y}$ is simply the mean of cluster totals.

$$\hat{Y} = \frac{A}{a}\sum_{i=1}^{a} Y_{i.} = \frac{A}{a}\sum_{i=1}^{a}\sum_{j=1}^{B_i} y_{ij} \quad (2)$$

The variance and the coefficient of variation of the estimator given in () respectively are defined as

$$V(\hat{Y}) = \frac{A(A-a)}{a}S_B^2, \quad S_B^2 = \frac{1}{A-1}\sum_{i=1}^{A}(Y_{i.}-\overline{Y})^2 \quad (3)$$

$$CV(\hat{Y})\% = \frac{\sqrt{V(\hat{Y})}}{Y}\times 100 = \left(\sqrt{\frac{A-a}{aA}}\right)\frac{S_B}{\overline{Y}}\times 100 \quad (4)$$

The quantity S_B^2 is the variance of enumeration area totals and measures the extent to which enumeration area total differs from one another. Both $V(\hat{Y})$ and $CV(\hat{Y})$ are measures of precision of the estimator given in (3 and 4). These quantities are useful in determining the number of sample enumeration areas to be selected that satisfies a targeted level of precision given assumed values of $k = S_B/\overline{Y}$. From the results of the COPD, values of k can be calculated based on test variables that

$$a = \frac{Ak^2}{A\delta^2 + k^2} \quad (5)$$

Where $k = S_B/\overline{Y}_B$ and δ desired precision target in terms of the coefficient of variation of the estimated total.

Using (5) and for different levels of targeted precision in terms of the coefficient of variation (est) set at 5%, 10%, 15%, and 20% and coefficient of variation values in Table A1.1, the calculated sample sizes under the assumption of simple random sampling are shown in Table A1.4.

Table A1.4: Calculated Sample Size (Number of Enumeration Areas) at Different Levels of Targeted Precision

Design – Test Variable	Sample Size (Number of EAs) for Given Precision Level [CV (est)]			
	5%	10%	15%	20%
Number of HHs with Agricultural Activity	50	27	15	9
Number of HHs Growing Crops	53	24	13	8

CV = coefficient of variation, EA = enumeration area, HHs = households.
Source: 2021 Census of Population and Dwellings, Cook Islands and author's calcuations.

The results in Table A1.4 can be used as a guide as to the level of the sample size for the PES. Ideally, a sample size that would generate PES estimates with the greatest precision in terms of a lower targeted coefficient of variation (est) is wanted. However, this needs to be balanced with available resources (personnel, time, and cost) and also utilizing the sample size for the PES in 2011. Note that in 2011, a sample of nine enumeration areas were selected for the PES. The choice of the sample enumeration areas was done purposively (i.e., the enumeration areas suspected to have large undercoverage). Based on available resources, it was decided to have a sample size of 12 enumeration areas. Such a sample is expected to yield a coefficient of variation (est) of 15%–20%. In particular, the expected coefficients of variation (est) with a sample of 12 enumeration areas were calculated using (4) as 15.8% for the number of households with agricultural activities and 17.2% for the number of households growing crops. With stratification, the final coefficient of variation (est) is expected to be lower than these computed values. The 12 sample enumeration areas will be allocated equally to the different strata and a sample of four enumeration areas will be selected with equal probability in each stratum. All households with agricultural activity identified by the 2022 COPD will be completely enumerated. Figure A1 shows the map of the sample enumeration areas selected for the 2022 PES.

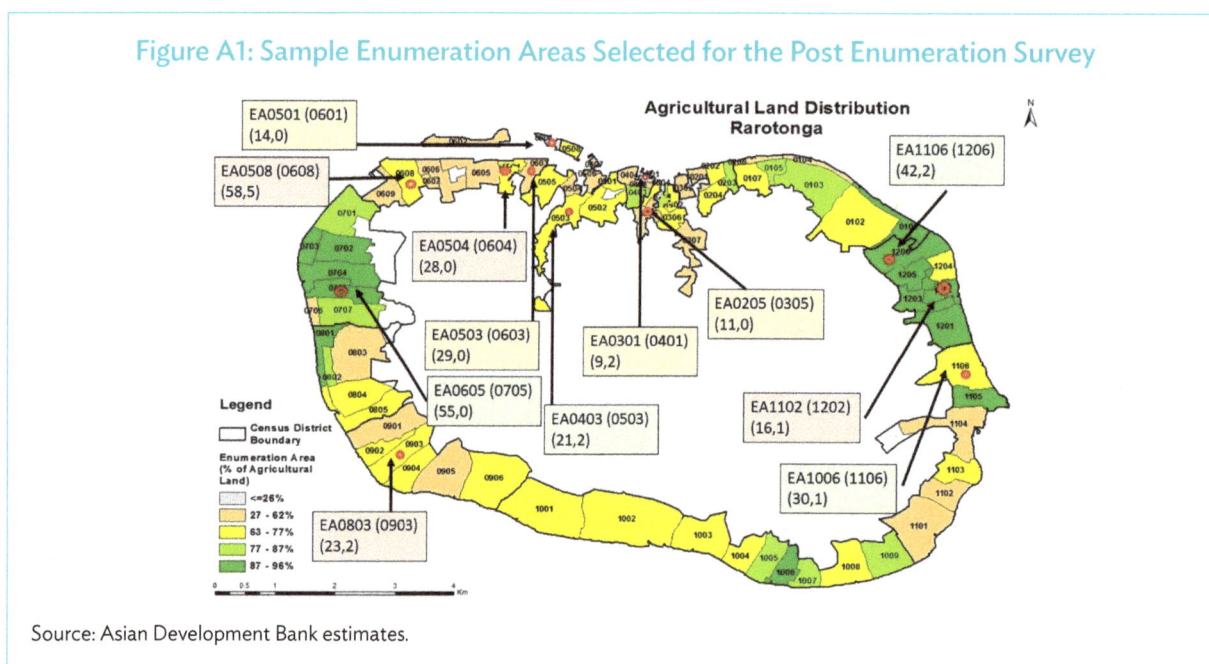

Figure A1: Sample Enumeration Areas Selected for the Post Enumeration Survey

Source: Asian Development Bank estimates.

Estimation Procedure

Surveys (including this PES) often involve the estimation of totals. Under a stratified one-stage cluster sampling design, an estimator of a total is given as

$$\hat{Y}_{st} = \sum_{h=1}^{L} \sum_{i=1}^{a_h} w_{hi} y_{hi.} \qquad (6)$$

Where the subscript h refers to stratum membership and i refers to the enumeration area in stratum h. Further, L is the number of strata (=3), and a_h is the number of sample enumeration areas in stratum h (=4). Also, y_{hi} and x_{hi} are the enumeration area totals (e.g., total number of households growing crops). In addition, w_{hi} is the final survey weight attached to all units in each enumeration area. Specifically, the final survey weight is defined as:

$$w_{hi} = \frac{A_h}{a_h} \times \frac{1}{R_{hi}} \qquad (7)$$

Where R_{hi} is the actual response rate in enumeration area i from stratum h.

As a measure of precision, the estimated variance of the total is given by:

$$s^2(\hat{Y}_{st}) = \sum_{h=1}^{L} A_h^2 \left(\frac{A_h - a_h}{a_h A_h} \right) s_{bh}^2,$$

(8)

$$s_{bh}^2 = \frac{1}{a_h - 1} \sum_{i=1}^{a_h} (y_{hi} - \overline{y}_h)^2, \quad \overline{y}_h = \frac{1}{a_h} \sum_{i-1}^{a_h} y_{hi}$$

Where s_{bh}^2 is the estimated variance between enumeration area totals and $\overline{y}_h$ is the sample mean of enumeration area totals.

Other useful measures of precision are the standard error which is defined as the positive root of the estimated variance, and the coefficient of variation (est) which is the ratio of the estimated standard error and estimated population total.

APPENDIX 2
DETAILS OF AREA SAMPLING FRAME FOR 2022 POST ENUMERATION SURVEY OF THE COOK ISLANDS

Datasets

(i) Land Use Land Cover (LULC) by European Space Agency (Raster was used in TIFF format)

   a. Worldview-2 (Resolution = 0.46m)
   b. Sentinel-2 (Resolution 10 - 60m)

(iii) Enumeration Area and Census District Boundaries (Data in Shapefile (shp) format provided by Team)

Methodology

The LULC classification included the following categories, the classes highlighted were included in the agricultural area estimation.

Table A2.1: Land Use Land Cover Classification

S.No	Class Code	Class Name
	A11	Broadleaved evergreen forest
	A12	Broadleaved shrubs
	A13	Grassland
	A31	Tree plantation
	A33_S-2	Terrestrial herbaceous vegetation – managed
	A33_WV-2	Terrestrial herbaceous vegetation – managed
	A33_WV-2_S-2	Terrestrial herbaceous vegetation – managed
	A43_S-2	Aquatic herbaceous vegetation – managed
	A43_WV-2	Aquatic herbaceous vegetation – managed
	A43_WV-2_S-2	Aquatic herbaceous vegetation – managed
	A63	Terrestrial herbaceous vegetation – management unknown
	A73	Aquatic herbaceous vegetation – management unknown
	B11	Bare soil
	B11A34_S-2	Temporarily bare managed land
15.	B11A34_WV-2	Temporarily bare managed land
16.	B11A34_WV-2_S-2	Temporarily bare managed land

continued on next page

Table A2.1 *continued*

S.No	Class Code	Class Name
17.	B12	Sand
18.	B21	Paved Surface/Rock
19.	B22	Building
20.	B3	Water Bodies
21.	C	Cloud/Cloud Shadow

An attribute table was computed for the raster image followed by reclassification of the raster in QGIS. The attribute table contains grid codes corresponding to each LULC class used for area estimation.

The raster was then converted to a shapefile (polygons) which were then dissolved to combine the polygons based on their grid codes.

The LULC shapefiles intersected with the enumeration area to extract the areas under the enumeration area boundary only.

The shapefile was then projected onto the WGS 84 / UTM Zone 4S (Projected Coordinate System) and the geometric area was calculated using the field calculator in QGIS.

The percentage of agricultural area was calculated by (Agricultural Area calculated from LULC/Total area of enumeration area *100).

Table A2.2: Sampling Frame (List + Area) for the 2022 Post Enumeration Survey

Variable Name	Variable Label
cendist	Census District Number
cendist_name	Census District Name
ceaid	Enumeration Area (EA) Code
nhh	Number of Households (HHs)
hhsize	Mean number of members per household
agri_ind	Number of HHs with agricultural activity as defined in the 2022 Census of Population and Dwelling (COPD)
fish_ind	Number of HHs with fishing activity as defined in the 2022 COPD
crop_ind	Number of HHs growing crops as defined in the 2022 COPD
live_ind	Number of HHs raising livestock as defined in the 2022 COPD
ea_area_ha	Estimated land area of the EA in hectares (from ESA satellite maps)
ag_area_ha	Estimated area of the EA in hectares that can potentially be classified as agricultural area
per_agarea	Estimated percentage of EA area potentially agricultural: ag_area_ha/ea_area_ha
no_bldgs	Number of buildings counted from satellite image and other sources

COPD = Census of Population and Dwelling, EA = enumeration area, ha = hectare, HH = household.

Table A2.3: Census District

Cendist	Cendist_name	Ceaid	2022 Census of Population and Dwelling (List Frame)						Area Frame			
			nhh	hhsize	agri_ind	fish_ind	crop_ind	live_ind	ea_area_ha	ag_area_ha	per_agarea	no_bldgs
1	PUE-TUPAPA-MARAERENGA	101	24	3.00	20	11	17	6	29.23	26.6	91.00	55
1	PUE-TUPAPA-MARAERENGA	102	96	3.56	91	45	81	31	86.72	65.59	75.63	160
1	PUE-TUPAPA-MARAERENGA	103	77	3.58	73	33	61	24	59.64	47.94	80.38	145
1	PUE-TUPAPA-MARAERENGA	104	13	3.54	10	6	5	1	10.12	5.68	56.13	32
1	PUE-TUPAPA-MARAERENGA	105	24	4.71	22	13	18	4	12.23	10.43	85.28	45
1	PUE-TUPAPA-MARAERENGA	106	12	2.58	12	4	9	1	6.89	4.25	61.68	29
1	PUE-TUPAPA-MARAERENGA	107	26	3.81	24	11	19	9	22.46	17.24	76.76	53
1	PUE-TUPAPA-MARAERENGA	108	22	4.09	20	11	15	8	8.11	3.35	41.31	57
1	PUE-TUPAPA-MARAERENGA	109	23	3.22	21	7	19	9	8.6	6.37	74.07	25
1	PUE-TUPAPA-MARAERENGA	110	15	3.53	14	4	14	2	9.2	7.77	84.46	45
1	PUE-TUPAPA-MARAERENGA	111	25	3.32	25	8	20	6	17.06	12.38	72.57	43
2	TAKUVAINE	201	15	2.93	14	5	10	6	3.49	3.04	87.11	19
2	TAKUVAINE	202	23	3.35	21	7	13	8	7.24	4.88	67.40	27
2	TAKUVAINE	203	12	4.75	12	2	8	2	8.68	5.03	57.95	28
2	TAKUVAINE	204	14	3.71	13	6	10	4	4.76	3.93	82.56	15
2	TAKUVAINE	205	19	4.53	18	9	11	8	16.62	10.32	62.09	34
2	TAKUVAINE	206	20	3.20	20	7	16	7	14.15	10.39	73.43	27
2	TAKUVAINE	207	32	3.69	29	12	20	11	31.15	13.85	44.46	47
3	TUTAKIMOA-TEOTUE	301	13	3.69	12	7	9	3	3.27	0.8	24.46	8
3	TUTAKIMOA-TEOTUE	302	7	5.14	5	4	4	0	3.4	2.03	59.71	34
3	TUTAKIMOA-TEOTUE	303	15	3.33	13	6	12	4	12.03	7.93	65.92	71
3	TUTAKIMOA-TEOTUE	304	10	3.90	9	6	8	4	8.17	4.66	57.04	24
4	AVATIU-RUATONGA	401	15	5.40	15	5	12	11	6.9	4.68	67.83	36
4	AVATIU-RUATONGA	402	40	4.18	35	26	30	26	29.9	20.41	68.26	65
4	AVATIU-RUATONGA	403	37	3.30	33	18	21	24	45.43	35.1	77.26	57
4	AVATIU-RUATONGA	404	22	2.55	18	11	11	3	7.4	3.76	50.81	29
4	AVATIU-RUATONGA	405	36	3.36	32	17	24	18	28.75	18.98	66.02	68
4	AVATIU-RUATONGA	406	24	3.58	18	9	10	14	8.73	4.46	51.09	35
4	AVATIU-RUATONGA	407	9	3.11	6	5	5	3	2.71	1.4	51.66	12

continued on next page

Table A2.3 *continued*

Cendist	Cendist_name	Ceaid	2022 Census of Population and Dwelling (List Frame)						Area Frame			
			nhh	hhsize	agri_ind	fish_ind	crop_ind	live_ind	ea_area_ha	ag_area_ha	per_agarea	no_bldgs
4	AVATIU-RUATONGA	408	14	3.43	14	3	10	8	7.12	5.2	73.03	29
5	NIKAO-PANAMA	501	15	2.80	15	5	14	4	4.96	1.31	26.41	31
5	NIKAO-PANAMA	502	31	3.06	26	10	21	4	22.18	8.63	38.91	83
5	NIKAO-PANAMA	503	36	3.47	34	16	29	14	15.32	6.24	40.73	44
5	NIKAO-PANAMA	504	36	4.03	34	13	28	16	17.11	11.08	64.76	57
5	NIKAO-PANAMA	505	33	3.48	32	13	28	12	30.42	17.41	57.23	63
5	NIKAO-PANAMA	506	32	3.28	32	8	28	9	27.12	12.62	46.53	58
5	NIKAO-PANAMA	507	25	3.28	21	12	16	7	5.53	2.58	46.65	45
5	NIKAO-PANAMA	508	75	3.69	71	33	58	12	32.2	20.74	64.41	119
5	NIKAO-PANAMA	509	41	3.56	35	21	29	11	23.28	14.4	61.86	59
6	RUAAU	601	51	2.78	45	24	38	13	47.5	39.61	83.39	81
6	RUAAU	602	30	3.07	29	10	23	12	38.83	36.63	94.33	58
6	RUAAU	603	8	3.50	7	4	6	1	16.25	14.92	91.82	92
6	RUAAU	604	70	3.47	63	39	54	32	45.42	41.51	91.39	109
6	RUAAU	605	64	3.89	62	23	55	25	33.88	32.77	96.72	79
6	RUAAU	606	17	3.88	16	11	9	7	7.71	4.59	59.53	37
6	RUAAU	607	57	3.28	57	21	48	18	42.13	34.75	82.48	97
7	AKAOA	701	31	3.84	29	16	23	13	21.74	20.46	94.11	74
7	AKAOA	702	16	3.19	14	10	12	5	9.63	7.99	82.97	41
7	AKAOA	703	39	2.85	38	6	29	17	67.61	39.74	58.78	76
7	AKAOA	704	42	3.14	41	18	27	18	41.82	28.92	69.15	133
7	AKAOA	705	25	4.08	25	13	19	10	38.25	27.2	71.11	66
8	MURIENUA	801	50	3.28	50	22	48	20	48.94	29.8	60.89	85
8	MURIENUA	802	29	3.10	28	9	24	11	19.96	14.37	71.99	46
8	MURIENUA	803	26	2.15	25	3	23	2	38.56	28.44	73.76	73
8	MURIENUA	804	30	3.07	27	11	24	9	33.09	22.14	66.91	57
8	MURIENUA	805	33	3.58	33	17	18	14	60.84	32.26	53.02	54
8	MURIENUA	806	30	3.47	29	15	21	12	76.54	57.28	74.84	43
9	TITIKAVEKA	901	66	2.80	63	28	54	22	95.07	67.49	70.99	

continued on next page

Table A2.3 *continued*

Cendist	Cendist_name	Ceaid	2022 Census of Population and Dwelling (List Frame)						Area Frame			
			nhh	hhsize	agri_ind	fish_ind	crop_ind	live_ind	ea_area_ha	ag_area_ha	per_agarea	no_bldgs
9	TITIKAVEKA	902	58	2.71	57	18	45	20	125.51	90.28	71.93	114
9	TITIKAVEKA	903	32	3.44	28	8	24	8	53.66	40.71	75.87	27
9	TITIKAVEKA	904	10	2.70	8	5	7	3	26.55	18.47	69.57	26
9	TITIKAVEKA	905	43	3.09	42	13	37	7	29.55	24.6	83.25	78
9	TITIKAVEKA	906	20	2.95	19	11	14	6	21.45	19.32	90.07	39
9	TITIKAVEKA	907	10	2.50	8	5	7	2	17.94	15.37	85.67	28
9	TITIKAVEKA	908	41	3.39	36	17	30	14	43.49	32.61	74.98	83
9	TITIKAVEKA	909	69	2.87	65	22	54	20	41.53	35.07	84.44	44
10	NGATANGIIA	1001	27	3.07	25	15	19	11	70.85	41.88	59.11	105
10	NGATANGIIA	1002	20	4.05	18	11	13	5	29.1	18.21	62.58	66
10	NGATANGIIA	1003	36	3.17	35	14	24	13	29.24	22.21	75.96	84
10	NGATANGIIA	1004	27	2.63	26	10	18	4	54.87	34.36	62.62	73
10	NGATANGIIA	1005	37	3.14	33	18	26	9	28.38	27.48	96.83	97
10	NGATANGIIA	1006	32	3.03	32	11	30	12	58.16	42.72	73.45	71
11	MATAVERA	1101	45	3.02	43	16	40	23	46.46	42.05	90.51	95
11	MATAVERA	1102	21	3.38	21	9	16	11	17.67	16.89	95.59	35
11	MATAVERA	1103	34	3.18	34	15	25	14	28.99	25.99	89.65	57
11	MATAVERA	1104	23	4.65	22	16	18	15	18.29	12.96	70.86	34
11	MATAVERA	1105	37	3.65	35	18	30	16	23.75	22.21	93.52	54
11	MATAVERA	1106	52	3.56	49	24	42	21	50.05	47.55	95.00	93

[1] Initial sample survey design writeup for the 2022 PES.
[2] Government of the Cook Islands, Ministry of Agriculture. 2021. Census of Agriculture Enumerators Manual. p. 3.
[3] Government of the Cook Islands, Ministry of Agriculture. 2014. Cook Islands 2011 Census of Agriculture. pp. 3–4.
[4] ESA Map Data Sources: (a) Contains modified Copernicus Sentinel data [2020-2021]: Worldview-2 [2021] Maxar Technologies; (b) Land Cover Mapping: Denise Dejon (GREENSPIN) Matuš Hrnčhar (World from space).
[5] Prepared by Mashal Riaz, consultant, and team member.

2022 COOK ISLANDS POST ENUMERATION SURVEY QUESTIONNAIRE

COOK ISLANDS
2022 POST-CENSUS OF AGRICULTURE ENUMERATION SURVEY
MINISTRY OF AGRICULTURE OF THE COOK ISLANDS

The information being collected under the Statistics Act 1966 will be kept strictly confidential and will be used for statistical purposes only.

FORM 2 - HOLDING

SECTION 0: GEOGRAPHY (pre-interview)

ID01. Census District									D	D
ID02. Enumeration Area						E	A	C	C
ID03. Holding No.							H	H	H
ID04. Holding ID		D	D	E	A	C	C	H	H	H
ID05. Address	TEXT									
ID06. Date of Interview	D	D	/	M	M	/	Y	Y	Y	Y

SECTION 1: HOLDING INFORMATION

1.1: RESPONDENT DETAILS

START, READ TO RESPONDENT: Greetings, my name is [Interviewer name] with the Ministry of Agriculture. The Ministry is conducting a survey in follow-up to the recent 2021 Agriculture Census to measure agricultural areas using GPS measurement tools.

Participation in the survey will require approximately one hour, and your consent for our team to walk across and measure land on your holding using GPS measuring devices. Any information collected will be confidential and protected by the Statistics Act of 1966.

1.1.1. Full name of respondent (involved in managing this holding during 2021)	TEXT
1.1.2. Sex of respondent	○ 1. Male ○ 2. Female ○ 9. Prefer not to say
1.1.3. Age of respondent (in complete years)	
1.1.4. Contact	TEXT

1.1.5. Do you, [1.1.1], consent to participate in this survey?	○ 1. Consent
	○ 2. Refuse
	○ 3. Reschedule,
	    Date: ....
	    Time: ....

1.1.6 GPS coordinate of main dwelling (taken at homestay entrance)	Record up to 5 decimals

1.2: RESPONDENT REPORTED TOTAL HOLDING AREA

1.2.1. What unit of measure will be used to report land areas?
    1 acre = 0.40 hectare
    1 hectare = 2.47 acres
	○ 1. Acres
	○ 2. Hectares
	○ 3. Square meters
	○ 9. Other, specify: ....

1.2.2a. Total area of the holding in whole [1.2.2]? (Whole units, Enter "0" here if area is less than area unit)	[Integer, whole unit]
1.2.2b. Fraction of an [1.2.2]? (Enter "0" if area is whole number)	[fraction]

SECTION 2: REPORTED PARCEL DETAILS

2.1.1. Parcel no.	2.1.2. Parcel description	2.1.3. Location of parcel	2.1.4. Number of separate plots (including fallow)	2.1.5. Area of parcel			2.1.6 Land tenure of parcel	2.1.7. Has this parcel been used for crop cultivation in 2021?	2.1.8. If Yes to 2.17, which crops were planted? (select all that apply)	2.1.9. If Yes to 2.17, method of planting (select one)
				Area Unit	Area in whole units	Area in fractional units				
01		○ 1. Within village (same island) ○ 2. Outside village (same island) ○ 3. Other Island					○ 1. Borrowed ○ 2. Leased land ○ 3. Occupation right ○ 4. Customary	○ 0. NO, fallow ○ 1. NO, other ag ○ 2. YES	Crop list	○ 1. Single row ○ 2. Double row ○ 3. Mixed crop ○ 4. Scattered
02										
03										

2.1.9. TOTAL AREA SUM

CONFIDENTIAL

FORM		OF	

COOK ISLANDS
2022 POST-CENSUS OF AGRICULTURE ENUMERATION SURVEY
MINISTRY OF AGRICULTURE OF THE COOK ISLANDS

The information being collected under the Statistics Act 1966 and will be kept strictly confidential and will be used for statistical purposes only.

FORM 2A - GPS AREA MEASUREMENT BY PARCEL

SECTION 1. PARCEL DETAILS

1.1 Parcel Identification

			D	D	E	A	C	H	H	H
1.1.1 Holding ID (copy from Form 1, Case Identification)			D	D	E	A	C	H	H	H
1.1.2 Parcel No. (copy from Form 1, Parcel roster)									I	D

1.2 Safety and environment conditions for GPS measurement

1.2.1. Can the perimeter of this parcel be walked and measured?	○ 0. NO ○ 1. YES
1.2.2. If No to 1.2.1., please specify why this parcel cannot be walked or measured?	○ 1. Parcel is more than 2 hours away (by any transport) ○ 2. Household refused ○ 3. Unsafe conditions ○ 9. Other, specify: ....
1.2.3. What is the density of tree cover over the parcel?	○ 1. Dense ○ 2. Light ○ 3. None
1.2.4. Weather at the time of measurement?	○ 1. Clear/sunny ○ 2. Partly cloudy ○ 3. Cloudy ○ 4. Rainy
1.2.5. Method of transport to the parcel?	○ 1. Walking ○ 2. Car/ motorcycle ○ 3. Boat ○ 4. Other, specify:...........................
1.2.6. Please note any weather and automobile safety issues when traveling to this parcel?	TEXT

1.2.7. Are health and automotive services readily accessible from the holding?	○ 0. NO ○ 1. YES
1.2.8. Other Notes	TEXT

1.3 Parcel characteristics

1.3.1. Sketch of parcel (Take note of the number of corners and visible markers [i.e., trees, structures])

1.3.2. Number of corners on parcel?	[integer]
1.3.3. Is the parcel on a slope?	○ 0. NO ○ 1. YES
1.3.4. If Yes to 1.2.3., based on your judgment, what is the slope of this parcel?	○ 1. Flat ○ 2. Walkable incline ○ 3. Climbable incline
1.3.5. Is the parcel fallow?	○ 0. NO ○ 1. YES
1.3.6. If No to 1.3.5., What is the main land use of this parcel?	Observable Land use categories
1.3.7. If No to 1.3.5., Is information being collected post harvest?	○ 0. NO ○ 1. YES
1.3.8. If No to 1.3.5., What is the primary crop grown	Crop list
1.3.9. If No to 1.3.5., Are there multiple crops (intercropping) in the parcel?	○ 0. NO ○ 1. YES
1.3.10 If No to 1.3.5., What percentage of the parcel is intercropped?	○ 1. 100% ○ 2. 75-100% ○ 3. 50-75% ○ 4. 25-50% ○ 5. Less than 25%
1.3.11. If Yes to 1.3.9., What is the secondary crop grown	Crop list
1.3.12. If Yes to 1.3.9., Is this information collected post-harvest?	○ 0. NO ○ 1. YES
1.3.13. In your estimate, is the parcel larger than 5x5 square meters?	○ 0. NO ○ 1. YES

SECTION 2. GPS MEASUREMENTS

2.1.1. Coordinate (from GPS) of <u>STARTING POSITION</u>										2.1.2. Photo to center	
Latitude				.						PHOTO	
Longitude				.							

2.1.3 File name of starting position	D	D	E	A	C	H	H	H	W	N	O

2.1.4. Parcel Area (from GPS) in acres					.		

2.1.5 File name of area track	D	D	E	A	C	H	H	H	A	N	O

PROTOCOLS FOR GPS PARCEL AREA MEASUREMENT

1. Inspect the parcel boundary. With the agricultural holder leading, you will first walk around the boundary of the parcel without the GPS device. Note any landmarks (i.e., trees, buildings, ponds, etc.) and overhead obstructions that may affect the accuracy of the GPS (e.g., dense forest canopy).

2. Sketch the parcel boundary. Sketch on Form 2A and outline of the parcel walked, clearly indicating corners and landmarks encountered along the way. Once complete, label the corners in order of the path to be walked and indicate the starting position in the northwest corner (Figure A4.1).

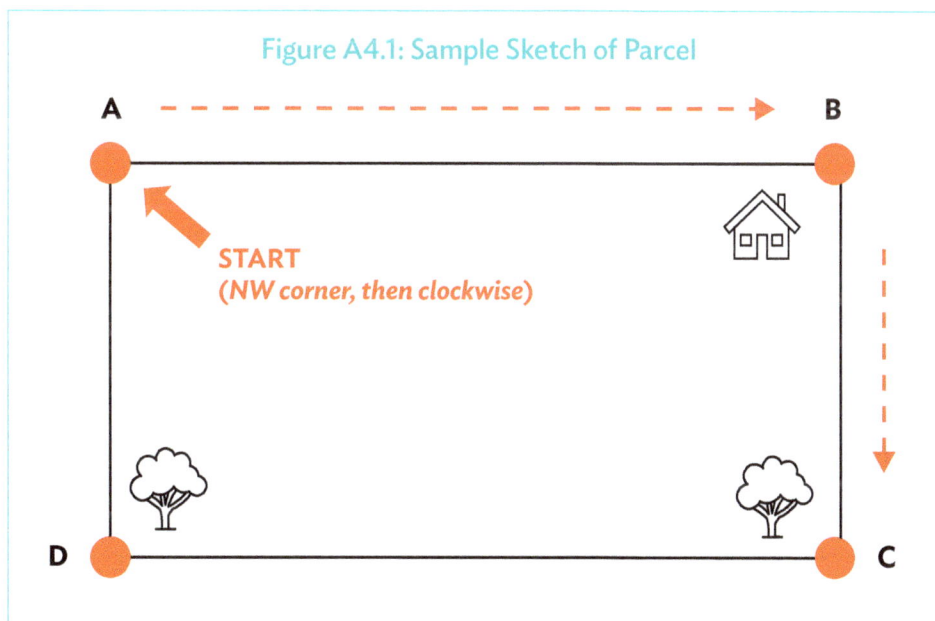

Figure A4.1: Sample Sketch of Parcel

START
(NW corner, then clockwise)

3. Parcel eligibility for GPS area measurement. If the parcel walked is less than 5 meters (m) x 5 m and it is unsafe to walk the boundary, then this parcel is ineligible for measurement. Proceed to the next parcel.

4. Prepare your equipment. Power on the GPS device and wait 60 seconds for the device to enter the main screen. Once powered on, navigate to the Satellite menu, and check your satellite accuracy (signal strength). If accuracy exceeds 5 m, continue to wait 60 seconds to allow the device to connect to additional satellites. If accuracy persists above 5 m for over 5 minutes, proceed with the GPS area measurement (Figure A4.2).

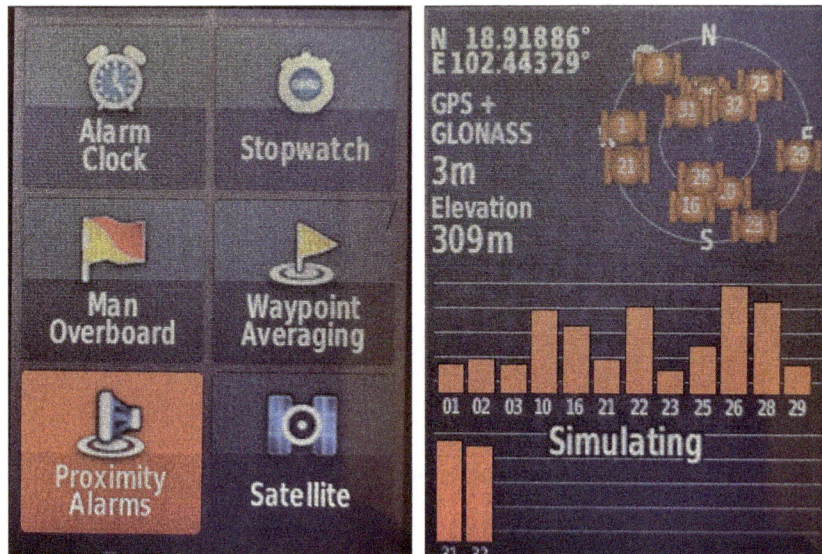

Figure A4.2: Checking Satellite Accuracy on the Garmin eTrex 32x

5. Travel to the starting position. Upon reaching the starting point, record a waypoint using the handheld GPS device using the steps in Figure A4.3. Enter the details of the recorded waypoint into the data collection application (format: S 12.32820 and E 125.33209), the time of the measurement, and proceed to take a photo toward the center of the parcel from the starting position. Before walking to the next corner, drop an object (i.e., a hat) to note the starting point on your return.

Figure A4.3: Marking a Waypoint on the Garmin eTrex 32x

1	2	3	4
Select the "Mark Waypoint" app from Main Menu	Change name to household ID to mark the household entrance or parcel ID to mark the parcel starting point	Log the lat/ long coordinate into the CAPI application	Click "Done" to complete the recording

6. Track the parcel using the Area Calculation feature. From the starting position, open the Area Calculation feature from the main menu on your GPS device. Once you are ready to walk the parcel, click "Start" and walk slowly with consistent pacing to the next corners of the parcel. At each corner, you should pause for at least 10 seconds before proceeding to the next. Upon returning to the starting position (indicated by your marker on the ground), click "Finish" on the GPS device to end tracking (Figure A4.4).

Figure A4.4. Marking a Waypoint on the Garmin eTrex 32x

1. Open "Area Calculation" app and select "Area Calaculation"

2. When ready, click "Start" to begin traversal

3. Proceed slowly next parcel corner, and pause at each corner for at least 10 seconds

4. Once you have returned to the starting point, click "Calculate"

7. Saving the area of the parcel. Reading first from the Garmin GPS device, enter the area calculated by the GPS device in acres up to 5 decimals. Next, on the GPS device, save the track using the naming convention [Holding number]-[Parcel number]a. The CAPI application will specify the filename (Figure A4.5).

Figure A4.5: Saving the Parcel Area

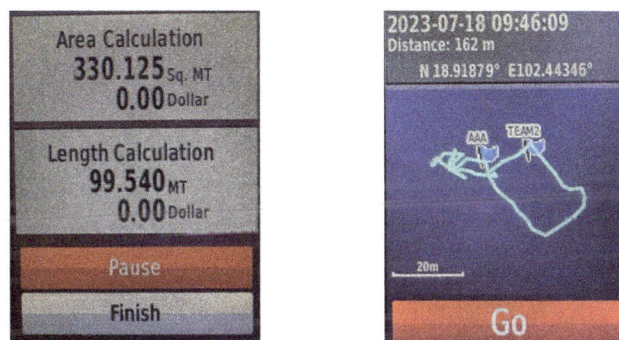

- Once finished, enter the displayed area into the CAPI application (acres with 3 decimals)

- Save the track, then select "Track Manager" in the main menu to check the parcel shape

Steps 1–7 are summarized for reference in Figure A4.6. All seven steps should be repeated for each parcel operated by the holding regardless if it is fallow or not.

Figure A4.6: Workflow of GPS Area Measurement Protocol

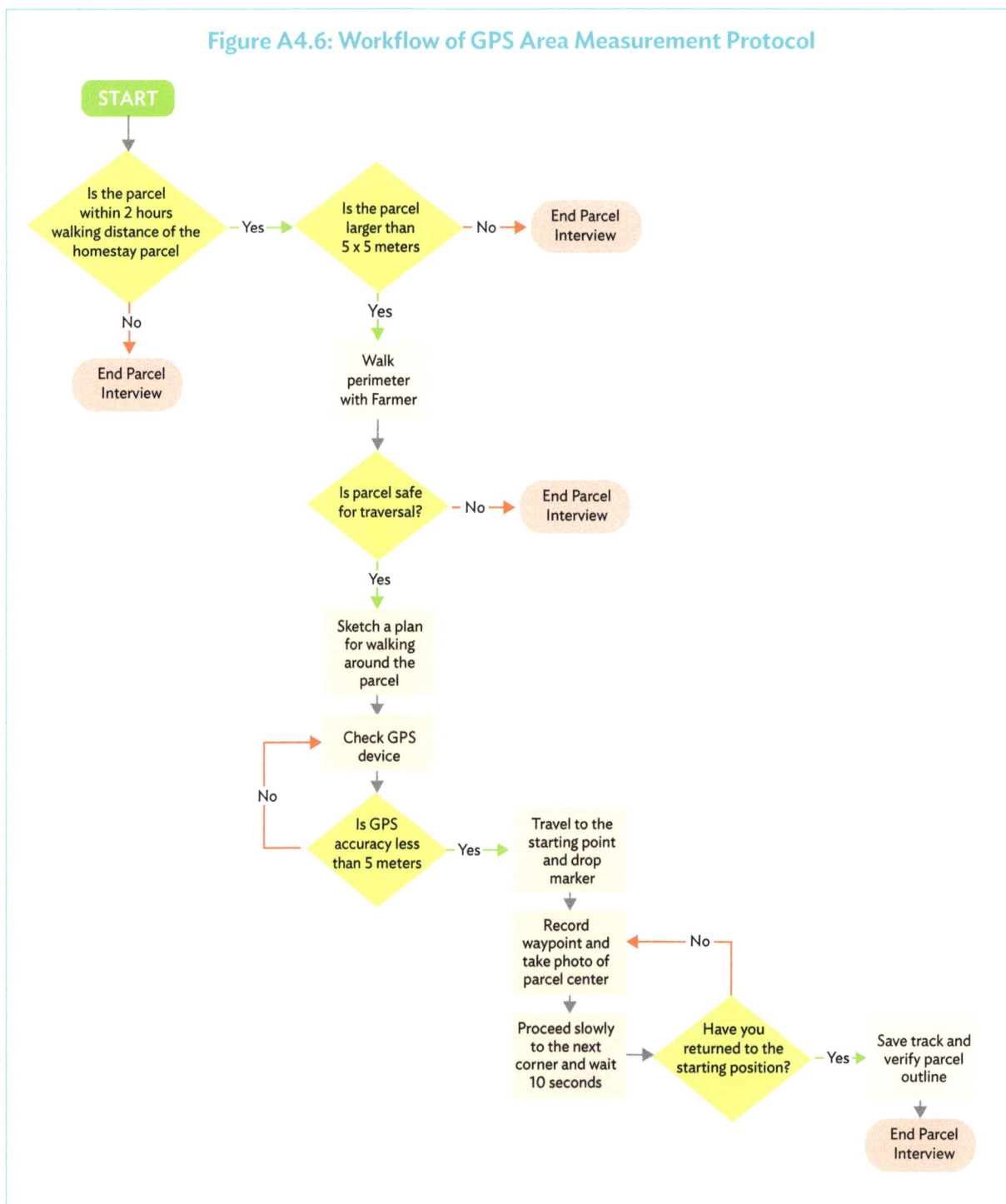

PROCEDURES FOR EDITING GPS DATA COLLECTED FOR AREA MEASUREMENT

Data management

1. Download gpx data files from handheld global positioning system (GPS) devices. If required, editing of the file names to uniquely identify the individual gpx files may be required.

2. Consolidate gpx files into a single file, or in batches according to the requirements of the analysis. For larger surveys, it may be optimal to organize the files by administrative area to keep the overall file size manageable for editing locally.

3. Verify the count and completeness of the points and parcel boundaries against the microdata set.

Cleaning GPS point data

1. Validate file names in the format specified during data collection. For the post enumeration survey (PES), these files should be named in the format of HouseholdID-ParcelID. This allows the analyst to reference each observation uniquely and to use a key to associate it with other tabular data files.

2. Remove duplicate points: Identify and (re)move any duplicate GPS points that have the same or very similar coordinates. This can be done by comparing the coordinates of each point with a specified tolerance value for latitude and longitude differences.

3. Remove outliers: Identify and remove any erroneous points or outliers that are significantly distant from the main polygon. This can be done manually by examining the data or tools in Quantum Geographic Information System (QGIS) utilizing the Local Outlier Factor.

4. Assess the accuracy of GPS points (optional): Check the accuracy of each GPS point based on the accuracy metric, and against other sources (i.e., intersect of points in each enumeration area).

5. Validate the cleaned GPS points: Compare the cleaned GPS points to the original raw data and any available reference data (e.g., satellite imagery, official maps) to assess their accuracy and quality.

Cleaning parcel polygon data

1. Validate file names in the format specified during data collection. For the PES, these files should be named in the format of HouseholdID-ParcelID. This allows the analyst to reference each observation uniquely and to use a key to associate it with other tabular data files.

2. Calculate the area of the raw import polygons. Using the $area function, calculate the areas of the raw imported polygons.

3. Import the polygons of the boundaries digitized by Survey Solutions and calculate the area similar to the previous step.

4. Calculate the difference between the areas of the farmer-reported estimate with the raw imported polygon file from the GPS and the difference between the farmer-reported estimate with the polygon delineated in Survey Solutions.

5. Split the polygons into quintiles based on the calculated difference. This will inform the priority list for data cleaning. Start with polygons with the largest observed distance.

6. Simplify the polygon shape. The Ramer-Douglas-Peucker (RDP) algorithm is used to simplify geometries, such as lines or polygons, by reducing the number of vertices while maintaining the overall shape.

7. Check for self-intersections: Check if the polygon does not contain any self-intersecting lines or overlapping sections. If self-intersections are present, manually edit the polygon or look into using an automated technique to resolve them.

8. Conduct manual adjustments: Using the digitized tools in Quantum Geographic Information System (QGIS) adjust the points of the polygon in line with reference data.

9. Validate the cleaned polygon: Compare the cleaned polygons to the original raw data and any available reference data (e.g., satellite imagery, official maps) to assess their accuracy and quality.

10. Export the cleaned polygon: Once the cleaned polygon meets your accuracy and quality requirements, export it to your desired file format.

REFERENCES

K. A. Abay et al. 2019. Correlated non-classical measurement errors, "Second best" policy inference, and the inverse size-productivity relationship in agriculture. *Journal of Development Economics*. 139. pp. 171–184.

K. A. Abay et al. 2022. Nonclassical Measurement Error and Farmers' Response to Information Reveal Behavioral Anomalies. Policy Research Working Paper. No. WPS9908. World Bank. https://documents.worldbank.org/en/publication/documents-reports/documentdetail/737051643050587043/Nonclassical-Measurement-Error-and-Farmers-Response-to-Information-Reveal-Behavioral-Anomalies.

G. Azzari et al. 2021. Understanding the Requirements for Surveys to Support Satellite-Based Crop Type Mapping: Evidence from Sub-Saharan Africa. Policy Research Working Paper. No. WPS9609. World Bank. https://documents.worldbank.org/en/publication/documents-reports/documentdetail/339781617301798195/Understanding-the-Requirements-for-Surveys-to-Support-Satellite-Based-Crop-Type-Mapping-Evidence-from-Sub-Saharan-Africa.

K. G. Beegle et al. 2015. *Reliability of Recall in Agricultural Data*. Brief No. 147749. World Bank. https://documents.worldbank.org/en/publication/documents-reports/documentdetail/862631587131538831/Reliability-of-Recall-in-Agricultural-Data.

C. Carletto et al. 2013. From guesstimates to GPStimates: Land area measurement and implications for agricultural analysis. *Policy Research Working Paper.* No. WPS6650. World Bank. https://documents.worldbank.org/en/publication/documents-reports/documentdetail/476181468327003343/From-guesstimates-to-GPStimates-land-area-measurement-and-implications-for-agricultural-analysis.

C. Carletto et al. 2015. *Measurement, Farm Size and Productivity.* Brief No. 147744. World Bank. https://documents.worldbank.org/en/publication/documents-reports/documentdetail/668871587125098246/Measurement-Farm-Size-and-Productivity.

C. Carletto et al. 2016-a. Cheaper, faster, and more than good enough: Is GPS the new gold standard in land area measurement? *Policy Research Working Paper.* No. WPS7759. World Bank. https://documents.worldbank.org/en/publication/documents-reports/documentdetail/257481469116364362/Cheaper-faster-and-more-than-good-enough-is-GPS-the-new-gold-standard-in-land-area-measurement.

C. Carletto et al. 2016-b. Land Area Measurement in Household Surveys: Empirical Evidence and Practical Guidance for Effective Data Collection. *Working Paper.* No. 147692. World Bank. https://documents.worldbank.org/en/publication/documents-reports/documentdetail/606691587036985925/Land-Area-Measurement-in-Household-Surveys-Empirical-Evidence-and-Practical-Guidance-for-Effective-Data-Collection.

C. Carletto et al. 2017. Mission impossible? Exploring the promise of multiple imputation for predicting missing GPS-based land area measures in household surveys. *Policy Research Working Paper.* No. WPS8138. World Bank. https://documents.worldbank.org/en/publication/documents-reports/documentdetail/668211499349698549/Mission-impossible-exploring-the-promise-of-multiple-imputation-for-predicting-missing-GPS-based-land-area-measures-in-household-surveys.

C. Carletto et al. 2021. Agricultural Data Collection to Minimize Measurement Error and Maximize Coverage (English). *Policy Research Working Paper.* No. WPS9745. World Bank. http://documents.worldbank.org/curated/en/751081627578468610/Agricultural-Data-Collection-to-Minimize-Measurement-Error-and-Maximize-Coverage.

S. Desiere and D. M. Jolliffe. 2017. Land productivity and plot size: Is measurement error driving the inverse relationship? *Policy Research Working Paper.* No. WPS8134. World Bank. https://documents.worldbank.org/en/publication/documents-reports/documentdetail/119781498874101671/Land-productivity-and-plot-size-is-measurement-error-driving-the-inverse-relationship.

A. S. Dillon et al. 2016. Land measurement bias and its empirical implications: Evidence from a validation exercise. Policy Research Working Paper. No. WPS7597. World Bank. https://documents.worldbank.org/en/publication/documents-reports/documentdetail/258851468197930768/Land-measurement-bias-and-its-empirical-implications-evidence-from-a-validation-exercise.

I. Yacoubou Djima and T. Kilic. 2021. Survey Measurement Errors and the Assessment of the Relationship between Yields and Inputs in Smallholder Farming Systems: Evidence from Mali. Policy Research Working Paper. No. WPS9841. World Bank. https://documents.worldbank.org/en/publication/documents-reports/documentdetail/711441636127459189/Survey-Measurement-Errors-and-the-Assessment-of-the-Relationship-between-Yields-and-Inputs-in-Smallholder-Farming-Systems-Evidence-from-Mali.

S. Gourlay and T. Kilic. 2022. Is Dirt Cheap? The Economic Costs of Failing to Meet Soil Health Requirements on Smallholder Farms. *Policy Research Working Paper.* No. WPS100108. World Bank. https://documents.worldbank.org/en/publication/documents-reports/documentdetail/099025206272226611/IDU0ee487ee70a84304c8c08b8a031ef5894a785.

S. Gourlay et al. 2017. Could the debate be over? Errors in farmer-reported production and their implications for the inverse scale-productivity relationship in Uganda. *Policy Research Working Paper.* No. WPS8192. World Bank. https://documents.worldbank.org/en/publication/documents-reports/documentdetail/242721505231101959/Could-the-debate-be-over-errors-in-farmer-reported-production-and-their-implications-for-the-inverse-scale-productivity-relationship-in-Uganda.

S. Gourlay et al. 2019. *A New Spin on an Old Debate: Errors in Farmer-Reported Production and their Implications for Inverse Scale-Productivity Relationship in Uganda.* Journal Article. No. 173981. World Bank. https://documents.worldbank.org/en/publication/documents-reports/documentdetail/099320507152229633/P1622920f10cc80e9098570dc9f95aec838.

T. Kilic et al. 2013. Missing(ness) in Action: Selectivity Bias in GPS-Based Land Area Measurements. Policy Research Working Paper. No. 6490. World Bank. http://hdl.handle.net/10986/15849.

D. B. Lobell et al. 2018. Eyes in the sky, boots on the ground: Assessing satellite- and ground-based approaches to crop yield measurement and analysis in Uganda. *Policy Research Working Paper.* No. WPS8374. World Bank. https://documents.worldbank.org/en/publication/documents-reports/.documentdetail/556261522069698373/Eyes-in-the-sky-boots-on-the-ground-assessing-satellite-and-ground-based-approaches-to-crop-yield-measurement-and-analysis-in-Uganda.

D. B. Lobell et al. 2022. Eyes in the Sky, Boots on the Ground: Assessing Satellite, and Ground-Based Approaches to Crop Yield Measurement and Analysis. Report No. 173982. World Bank. https://documents.worldbank.org/en/publication/documents-reports/documentdetail/099320407152277828/P16229209200bf0bd0b8820e733ebdb3407.

D. C. Miller et al. 2019. Trees on Farms: Measuring Their Contribution to Household Welfare. World Bank. https://www.worldbank.org/en/programs/lsms/publication/Trees-on-farms-measuring-their-contribution-to-household-welfare.

H. Park et al. 2022. Geography, Institutions, and Global Cropland Dynamics. Policy Research Working Paper. No. WPS10078. World Bank. https://documents.worldbank.org/en/publication/documents-reports/documentdetail/099959306092221155/IDU05dfecdd70a76e046370b1be0d76986660e6d.

T. G. Tiecke et al. 2017. Mapping the World Population One Building at a Time. Working Paper. No. 148052. World Bank. https://documents.worldbank.org/en/publication/documents-reports/documentdetail/439381588065763562/Mapping-the-World-Population-One-Building-at-a-Time.

J. Tomaštík Jr. et al. 2017. Horizontal accuracy and applicability of smartphone GNSS positioning in forests. Forestry: An International Journal of Forest Research. 90(2). pp. 187–198. https://doi.org/10.1093/forestry/cpw031.

www.ingramcontent.com/pod-product-compliance
Lightning Source LLC
Chambersburg PA
CBHW061221270326
41926CB00032B/4812